商务英语口语实训教材

第 4 册

高新技术、技术转让与国际工程合作

主　编	冉隆德	李　才	刘安洪	罗玲华	强亚丽
副主编	陈严春	周　茜	初兴春	段其为	邹林凤	周湘婷
编　委	惠　鑫	单宇鑫	戚萍萍	洪文静	魏敏俐	姜　俊
	李韶希	汤卫平
顾　问	梁瑞雄
主　审	汤义贤	李弻苍（美籍）

暨南大学出版社
中国·广州

图书在版编目（CIP）数据

高新技术、技术转让与国际工程合作/冉隆德等主编. —广州：暨南大学出版社，2013.3
（商务英语口语实训教材）
ISBN 978-7-5668-0404-4

Ⅰ. ①高… Ⅱ. ①冉… Ⅲ. ①高新技术—国际科技合作—研究 Ⅳ. ①TB②G321.5

中国版本图书馆 CIP 数据核字（2012）第 260454 号

出版发行：暨南大学出版社

地　　址：	中国广州暨南大学
电　　话：	总编室（8620）85221601
	营销部（8620）85225284　85228291　85228292（邮购）
传　　真：	（8620）85221583（办公室）　85223774（营销部）
邮　　编：	510630
网　　址：	http：//www.jnupress.com　http：//press.jnu.edu.cn
排　　版：	广州市天河星辰文化发展部照排中心
印　　刷：	广州市怡升印刷有限公司
开　　本：	787mm×1092mm　1/16
印　　张：	8.25
字　　数：	190 千
版　　次：	2013 年 3 月第 1 版
印　　次：	2013 年 3 月第 1 次
定　　价：	22.00 元

（暨大版图书如有印装质量问题，请与出版社总编室联系调换）

Preface
序 言

21 世纪是科技与经济快速发展的世纪。科技不断创新与经济全球化是推动社会发展的两大动力。经济全球化和我国加入 WTO；科学技术的接力性和拓展性；新经济理论、制度、模式的创建；国际交往与合作在更多的国家之间、在更广泛的领域迅速发展和深入，都赋予了外语教学和外语人才新的内涵，并对其提出了更高的要求。

学生坚实的外语基础知识的形成，除了学生本人的智慧和勤奋学习外，我国外语教育的指导思想和实践，广大外语教师的思想观念，对学生学习与获得知识以及适应时代的要求，起着极大的导向作用。我们认为：外语教学的根本任务和永恒的主题是主动适应和促进国家的经济、科技、社会发展，提高国民的素质；加强国际交往与合作，吸收、消化、升华和创新人类一切智慧结晶和优秀成果等。外语教学和研究如果偏离了这一主题，就会走弯路，甚至会误入歧途。

随着经济全球化的不断深入与发展，英语口语教学、学生口头交流的能力越来越重要。国际间的各类经贸活动、商务活动、商务谈判、国际工程合作、投资融资、三资企业、合同与协议的签署和实施、国际交往、各国高层互访、中欧、中非、亚太经合组织、上海合作组织、东盟经济圈、一系列高峰论坛、国际博览会、展销会等，从语言交流角度讲，主要是通过外语口语或口译来实现的。

特别是由于我国的经济实力不断增强，中国的国际地位不断提升，在世界的影响力日益强劲和扩大，中国走向世界、融入世界的步伐不断加快。数以万计的企业和公司在国外做生意，从事经济技术合作、科技合作、承包工程和基础设施项目，并购企业，进行农业合作、交通和通讯合作，在世界各国开展舞台巡演等文化交流，所有这些活动都需要大量精通业务，熟悉国际规则和国际商法，外语基础扎实，具有流利英语和其他外语口语能力的国际型人才。

随着中国改革开放的不断深入，数以万计的三资企业已在神州大地生根、开花结果。国外一些高新技术产业和研究开发中心正陆续进入中国，服务贸易将大大增加，这就急需既懂经济、科技、管理、金融、旅游专业知识，又能用专业、流利的英语或其他外语进行交流的人才。

综合性大学，理、工、农、医、财经、师范类院系和各类公办、民办职业技术院校，是我国科学家、工程技术专家、经济学家、高级技师、各类大师、建设者和领导人的摇篮。这些高校学子和留学归国学子是经济全球化，我国履行 WTO 责任，促进国际交流、合作、和谐的参与者、实践者和创新者，是我国快步走向世界、融入世界、加快自主创新和知识创新、提高自主研发能力的中坚力量，是我国借鉴世界各国优秀科技成果、智慧精华并加以升华，为我所用的优秀人才或领军人物。他们的外语素质和英语口语自主交流能力将在很大程度上决定着这些方面的进程和结果。

这些高校学子和留学回国的学子，是将来的董事长、总经理、首席执行官和各类公司、企业的员工。在对外交流与合作中，如果他们能用外语自如地进行交流，这不仅可以赢得竞争对手或合作伙伴的赞赏和尊重，克服语言文化障碍，而且还可以提升这些公司和企业的国际形象和信誉。这些学子中的部分人，可能是将来的部长、省长、市长、县长、镇长，在对外交流与合作中，如果他们能讲一口流利的英语口语，或能自如地用英语与国外友人、合作者交流，不仅可以得到对方的赏识和敬佩，而且可以在很大程度上提高这些部、省、市、县、镇的国际知名度并树立良好形象。

　　众所周知，科技、经济相对落后的国家向发达国家学习先进技术和管理理念，主要是通过语言自主交流实现的；创新能力和创新意识相对落后的国家向发达国家学习也主要是通过语言自主交流实现的；中方人士在国际机构和组织行使权利，主要是通过发言权实施的；外交、国际经贸、跨文化交流等也主要是通过外语自主交流实现的。因此，外语自主交流的功底和专业知识的广度、深度，决定着吸取他国人民智慧结晶和科技、经济优秀成果的广度、深度和准确度，影响着创新思维、创新成果的速度和创新人才的品质。

　　科技创新人才具有丰富的、独特的想象力，有超凡的逻辑思维、形象思维、创新思维和理念思维。各种科学想象和思维要在同行科学家、相关学科、交叉学科和实践中碰撞、论证，或被其他科学想象和思维所代替，使其深化、升华，更接近实际和真理，这有助于取得重大科技突破和成果。而这些科技创新人才要经常进行学术交流，必须有很强的科技外语自主沟通能力。

　　在当今社会的各种交流中，面对面、视频上的口头自主交流越来越频繁，涉及的领域十分广泛，特别重要的一点是，这种口头自主交流，有助于对科技疑难或前沿课题的深入探讨；有利于认清事物的本质，有益于及时并准确了解各方思想中的闪光点；有助于掌握各方创新思维中的苗头或核心。这是科技、经济相对落后的国家培养顶尖人才并缩小与发达国家科技和经济差距的一种快速方法。因此，在我国理、工、农、医和财经类院校开设内容与其学科、专业相适应的口语课是外语教改中一项迫在眉睫的重要任务。

　　高校的外语教学应高瞻远瞩，放眼世界，清醒地看到世界政治、经济、科技等形势的巨大变化。中国要在全球化的世界经济中富有竞争力，必须拥有国有和民营的各行各业的跨国公司。跨国公司是全球化的先锋和主力，中国最缺的是这类公司；中国要有能力驾驭全球化，必须拥有大批具有国际观、业务精、外语好、熟知国际商法的高端人才，中国最缺的就是这类人才。外语教学的重心就是要为培养国际型人才贡献自己的一份力量。因此，外语教学从内容到方法必须打破传统模式，要造就千千万万适应国际和国内政治、经济、科技、金融、管理、交往与合作、跨文化交流等领域的人才，必须强调高校外语口语教学的地位和重大意义，在改革开放的珠三角、长三角、天津滨海区以及其他沿海和内陆大中城市中的高校更应如此。

　　新中国成立以来，特别是改革开放以来，国家和各类高校十分重视外语教学，每年投入了大量人力、物力和财力，学生也投入了大量时间、金钱和精力学外语，但效果一直不尽如人意。一些专家学者惊呼："我国高教中最失败的是外语和体育！"有的领导也指出，"近20多年的'大学英语'教学，学生学成了'哑巴英语'"；"我国外语教学费时较多，收效较低，学生不满意，用人单位不满意，家长也不满意"。对这些评价和批评，外语界有些专家和学者可能持有异议，但我国高校学生外语口语表达不佳是一个不争的事实。

就"哑巴英语"而言，主要指的可能是：外语院系的学生能用英语进行礼仪接待、日常会话，而听不（大）懂、不（大）会说经贸、科技、国际交往与合作、金融、管理等专业英语；理、工、农、医和财经类等院系的学生，除了备考"大学英语"四、六级外，既不（大）会说生活用语、礼仪接待英语，也不（大）会用英语与同行交流科技、经贸、管理等方面的知识。其原因大致有以下几方面：①长期以来，各类高校不够重视学生口语训练；②某些外语、翻译类院系把英语口语学习主要局限在生活用语、礼仪接待内容方面；③外语口语教材和教学有两大弱点，一是没有突破日常会话、礼仪接待内容；二是试图突破日常会话、礼仪接待的外语口语教材，其中提及的商务、经贸、科技、管理等知识如蜻蜓点水，十分零散，学生无法学到较系统、知识面广的专业口语知识。④一些学者把学生的外语基础知识局限在外语语言体系本身及人文外语基础知识，长期忽视用外语表述的经贸、科技、管理、国际交往与合作、跨文化交流等方方面面的外语基础知识。⑤某些外语教材，一本书十多个单元，每单元有十七八个练习，却很少有口语练习，影响了学生学习更高层次、更广博的外语及口语知识，导致学生知识面窄，适应能力差。⑥邓小平同志指出，高等教育的关键，一是教师，一是教材。他精辟地阐释了学生学什么比如何学更重要的教学原则，并画龙点睛地说明了教师的关键作用。目前，我国外语教学在把握和执行这一原则中偏差颇大。

有鉴于此，我们在一定的基础之上，花了一年半左右的时间，组织编著了一套《商务英语口语实训教材》，共五册：①《商务礼仪接待与外贸基础会话》，主要内容有：建立商务关系；在广州国际会展中心的广交会；公司招待；商务宴请；理查到达伦敦；亨利在美国等51个会话。②《商务谈判与国际贸易》，主要内容有：商务谈判；洽谈电风扇的价格；艰难的谈判；我们应该努力消除分歧；国际贸易；进出口商务谈判；国际商务；补偿贸易；服务贸易等67个会话。③《市场营销、公司管理与投资融资》，主要内容有：公司在海外如何获利；工商企业的市场营销；汽车展馆；跨国公司；公司的国际竞争力；合资企业；品牌与诚信；投资：发展经济的发动机；在国外投资；项目融资等80个会话。④《高新技术、技术转让与国际工程合作》，主要内容有：电子化生活；数字技术；计算机模拟；智能机器；世界转基因农作物的发展和前景；高新技术展览；柔性制造技术；技术转让；国际工程项目等39个会话。⑤《商务合同、国际商法与WTO》，主要内容有：商务合同；国际商务的法律方面；公司法；美国的税务会计；WTO协议；WTO反倾销协议等35个会话。

我们编著此教材的理念：高校学生只有通过学习较系统、知识面涵盖较广的商务英语口语知识，结合本单位具体业务的英语知识，不断丰富、更新，才能适应经济全球化条件下的国际贸易、国际交往与合作等；突出篇章，未设练习，重在实训；商务英语、经贸、管理类专业以及外语、翻译院系的学生应把重点放在商务英语口语的学习和训练上，这是因为各类商务谈判、商务活动等，从语言交流角度讲，主要是通过口语、口译来实现的。

本书的特点：商务英语口语知识系统，由浅入深、涵盖面广、语言地道、知识丰富、内容新颖、实用性强。每册书后还有词汇拓展、语句展示、单证附录。本书以模块化方式编著，不同类型的商务英语内容相对集中，学生（员）可根据学习和工作的实际需要，五册书全学，或选其中两三册学习，或在每册书中选数篇会话重点学、突击学。

  本套书可用作大学经贸类、管理类、理工类等专业，外语院系、翻译院系的本科生、研究生的口语、口译课教材；也可作为各类职业技术院校、民办高校、成人教育学院、社会外语培训机构的口语实训教材；适用于英语翻译工作者、经贸人士、涉外企业营销人员、三资企业或涉外机构从业人员、相关公务员、金融机构职员、涉外律师、高级技师等使用；同时也是自学英语口语人士的良师益友。

  编写组成员在承担繁重的教学任务的同时，挤出宝贵的时间，全身心投入，充分利用寒暑假等节假日，克服重重困难，认真查阅，收集资料，一丝不苟地进行编著，反复校对，成功编著了知识较系统、内容较丰富、注释较翔实、话题较集中、形式较实用的《商务英语口语实训教材》。书本凝聚了编著者的辛勤劳动和智慧。其中冉隆德教授［1967年毕业于四川大学外语系英语专业，在中国科学院盐湖所工作21年，翻译了100余万字的科技文献资料；在国内外中英版杂志上发表译文、论文70余篇；接待过联合国及美国、英国、澳大利亚、新西兰、加拿大、德国、日本、瑞士等国的50多位科学家和学者，为其中30多人次的学术报告及两次国际学术会议担任口译，1985年任中国科学院硫酸钾考察团成员兼翻译访澳；1981—1987年曾在2003年国家最高科学奖获得者刘东生院士领导的中科院与澳大利亚第四纪合作项目中任翻译，受到刘院士的好评。还参加过4个中美等合作科研项目（任翻译）；1989年获中科院自然科学三等奖，1991年获青海省科技进步二等奖；1981年担任美澳阿尼玛卿山登山旅游队翻译。1987—2008年在重庆工商大学从事中层服务、管理、教学、科研21年，编著8本书，其中5本为教材，其教学经验丰富，科技、经贸知识比较全面］担任本书的总策划，搜集、查阅资料，组织编写，精心指导编写组每个成员注释会话，进行认真仔细的校对。参加编著工作的还有：广东科技学院的李才教授（1964年7月生，现任广东科技学院应用英语系主任，发表论文15篇，主持国家课题、子课题两项，省级课题一项，主编教材五本），刘安洪副教授（重庆文理学院外国语学院院长），卢兆强副教授、徐更生副教授、初兴春（讲师、教务处副处长）、陈严春（讲师、广东科技学院应用英语系副主任）；罗玲华（暨南大学硕士）、胡菊花（上海外国语大学硕士）、邹林风（湘潭大学硕士）、单宇鑫（香港浸会大学硕士）、梁树芬（中国地质大学外国语学院硕士），王苇（澳大利亚墨尔本英语教育硕士）等。

  本教材还是校企、校校结合的产物。参加本书编著工作的还有重庆市外贸局国际商务师余世民先生（从事进出口贸易30余年，1967年四川大学外语系英语专业毕业），他对教材内容编写提出了宝贵建议；重庆工商大学外语学院院长李文英教授、外事处处长周茜副教授等。

  本书由汤义贤教授（1968年毕业于武汉大学，长期从事高等教育工作，曾分别担任长江大学、广东科技学院外语系主任，教学经验丰富，科研成果丰硕）、李弼苍先生（美籍，在美国从事商贸工作近30年，长期奔忙于美国、香港、台湾、大陆，商务实践经验丰富，精通商务英语）担任主审；由广东科技学院党委书记梁瑞雄担任顾问。

  在本书编写过程中，我们得到广东科技学院领导、教务处、科研处的关心和支持。

  此外，本书还参考了一些编著者的劳动成果和智慧结晶，使本书内容更加丰富，知识领域更加宽广，话题更加全面，书的质量得到提高，我们在此向这些参考资料提及的所有编著专家、教授、学者和出版社表示崇高的敬意和衷心的感谢！

　　由于工作条件，编著者的水平、知识和经验有限，本书编排的科学性、分类的准确性、难易梯度及搭配不尽如人意，编著中的错误在所难免，欢迎专家、教授、学者、同行及广大读者赐教。我们对此表示衷心感谢！

《商务英语口语实训教材》编写组

2012 年 12 月 18 日于广东

Contents
目 录

Preface 序言 ·· 1

Part One　High-tech 高新技术

1. How to Use a Computer（怎样使用计算机）································· 2
2. Multimedia Technology（多媒体技术）··· 3
3. Internet（英特网）··· 5
4. Internet-based Training（基于英特网的培训）······································· 8
5. Hot Technologies of Computer（计算机热门技术）······························· 9
6. E-life（电子化生活）··· 11
7. E-commerce（电子商务）··· 14
8. Office Automation（办公室自动化）··· 17
9. Global Positioning System（GPS）（全球定位系统）······························ 18
10. Digital Technology（数字技术）··· 20
11. Automation & IT Application（自动化和信息技术的应用）····················· 22
12. Computer Simulation（计算机模拟）··· 24
13. Intelligent Machines（智能机器）··· 27
14. Robot and Its Recent Development（机器人及其近期的发展）················· 29
15. The FMS—An Advanced Manufacturing Technique（柔性制造技术———一种先进的制造技术）·· 33
16. Discussion of Manufacturing Technological Problems（1）（讨论制造技术问题之一）·· 38
17. Discussion of Manufacturing Technological Problems（2）（讨论制造技术问题之二）·· 41
18. Environmentally Friendly Farming in the U.S.A.（美国的环境友好型农业）············ 45
19. Cloning（克隆）·· 47
20. The Development and Prospect of Genetically Modified Organism Crops in the World（世界转基因农作物的发展和前景）·· 52
21. Zhuhai International Aviation and Aerospace Exhibition（ZIAAE）（珠海国际航空航天展）·· 54
22. Innovations in Business（企业创新）··· 59
23. China International High-tech Exhibition（中国国际高新技术展览会）········ 61

Part Two Technology Transfer 技术转让

24. Technology Transfer (1)（技术转让之一） ·· 65
25. Technology Transfer (2)（技术转让之二） ·· 68
26. Technology Transfer (3)（技术转让之三） ·· 70
27. Technology Transfer (4)（技术转让之四） ·· 72
28. The Royalty Rate and the Initial Down Payment Are Too High（提成费和入门费太高） ··· 75
29. Buying Know-how（购买专门技术） ·· 77
30. Transfer the Right to Use the Patent（转让专利使用权） ································ 78

Part Three International Engineering Project Cooperation 国际工程项目合作

31. International Engineering Project Contract and Service Cooperation (1)（国际工程项目合同及服务合作之一） ··· 82
32. International Engineering Project Contract and Service Cooperation (2)（国际工程项目合同及服务合作之二） ··· 84
33. International Engineering Project Contract and Service Cooperation (3)（国际工程项目合同及服务合作之三） ··· 87
34. A Turn-key Project（总承包工程项目） ··· 90

Part Four Inviting Bids and Bidding 招标投标

35. Inviting Bids and Bidding 招标投标 ·· 94
36. Different Opinions about Bid Invitation（关于招标的不同意见） ······················· 95
37. To Find a Bidder Who can Guarantee a Market（寻求有市场保证的投标商） ········· 97
38. It is Fairy Normal at This Stage for Bidder to Offer Extras（在投标的这个阶段投标商提供更优惠的条件很正常） ··· 99
39. Your Bid Has Been Accepted（你方中标了） ·· 101

Part Five Appendices 附录

1. Expanding Words and Phrases（词汇拓展） ·· 105
2. Drills of Sentences（句式展示） ·· 110
3. Appendices of Business Documents（商务单证附录） ································· 113

References 参考文献 ·· 120

Part One ①

High-tech

高新技术

1. How to Use a Computer
（怎样使用计算机）

James (J) decides to learn how to use a computer, and goes to the computer room in the college. He asks his classmate (C) there how to use the computer.

J: Excuse me, can you help me?

C: Oh, sure. What's the problem?

J: Well, I've never used a computer before, I am afraid that I have very little idea about what to do.

C: OK, well, you'd better start at the beginning, then... first you need to switch it on (He switches on the computer). Now, we've just got to wait for a minute while the computer is checking itself out... now just wait. Right, here it is, you see this colored screen, this is Windows 2000.

J: What is Windows 2000?

C: Windows 2000 is an operating system, which controls the hardware, reading from and writing to the disk, sending information to the printer and so forth. Now move the mouse. You see the arrow on the screen? It is controlled by the mouse. Move the arrow to the "Start" icon at the bottom of the screen and click the left-hand button of your mouse and you will get a menu.

J: What does the menu mean?

C: The menu is just the list of programs that are on the computer and you can choose the one you want.

J: Well, I just want to do some typing, type my essay, so what do I want actually?

C: Right, you need the Word Processor, that is, the "Microsoft Word" in this computer. Point the cursor to "Program" and you can find some of the programs that come with Windows 2000, that's it, OK. To start the typing program, click "Microsoft Word" now... There you are, blank screen. You could just start typing now, as if it's a typewriter.

J: Er... what if I make mistakes, I know it'll correct them, but how do I do that?

C: Mm, well, the mouse you used before, you can use it to move the cursor to the front of a mistake and click, then press "Delete", and type in whatever you want to correct it to. What's important is to save your document, then you can go away, come back to it... In order to save your document, you should use your mouse to click the "File" menu and find "Save" function there. Now click the "Save" and see what happens. Now you can see it asks you for a name, so now you need to type in a name you want to call your document.

J: OK, that if I call it by my name, J?

C: Right, and press "Enter" again, and there you can see it's saving for you.

J: OK, it seems quite easy actually.

C: The other things you need to know are, how to finish and how to exit. I'll tell you later. I need to go and get on with my work over there, and let me know when you've finished, and I'll come and help you exit.

J: Oh, thank you very much, you've been very helpful and kind. Thank you. Will you be here actually?

C: Yeah, I'm going to be here for the next couple of hours, so, sure.

J: So, if I get stuck, can you help me?

C: Yes, of course.

J: Thanks a lot.

Notes

1. The computer is checking itself out. 计算机在自动检索。
2. Move the arrow to the "Start" icon at the bottom of the screen. 移动箭头到屏幕下方的"开始"键。
3. Click the left-hand button of your mouse. 单击鼠标左键。
4. What do I want actually? 我该选哪个程序？
5. the Word Processor 文字处理器（程序）
6. the Microsoft Word 微软 Word 软件
7. Point the cursor to "Program". 将光标移到"程序"上。
8. blank screen 空白屏幕
9. "Save" function "存盘"程序
10. get stuck 计算机卡住了，死机

2. Multimedia Technology
（多媒体技术）

A: What is multimedia?

B: Well, it typically refers to a synthesis of graphics, animation, optical storage, image processing, and sound. It is not a single technology, product, or market. Instead, it is a collection of technologies that are in the process of being joined together.

A: It seems multimedia is something that makes something else more efficient. Can you give an example?

B: Yes. Suppose you have a list of written directions of getting to the bank and a mapping application that you could use to pinpoint where you want to go. Not only could that application print out detailed instructions of how to get to the bank, it could also display a full-color image of a map to the bank, along with a brief audio segment telling you the traffic conditions of roads around the bank.

A: Oh I see. This mapping application is multimedia in action, which is the combination of text (the printed instruction), static graphic images (the color map), and digital sound (the narration describing traffic conditions). It makes you get to the bank faster.

B: You are right. Multimedia has started to play an increasingly important role in today's computer world and is truly changing the way people use computers because of even more powerful computer systems and experiences of creative programmers.

A: What are the uses of multimedia in certain areas?

B: It can be used for computer-based training (CBT). Many companies are turning to multimedia application to train their employees. Then in education, it makes the learning process more interesting. There is a new type of software category—edutainment that mixes education with entertainment. And there is no doubt that it is of great use for entertainment. And you know, as we are in an age of information, multimedia provides effective ways to organize information and search for specific facts quickly and efficiently. In another area, multimedia is applied in business presentations.

A: Obviously, you have much to say, but I think I've got the idea. Multimedia has many applications and uses, and the only limitation is our imagination. Thank you very much.

B: You're welcome.

Notes

1. multimedia 多媒体
2. It typically refers to a synthesis of graphics, animation, optical storage, image processing, and sound. 它（多媒体）一般是指图形、动画、光存储、图像处理和声音的合成。
3. mapping application 地图应用程序
4. pinpoint 找到，精确地找到……的位置
5. a brief audio segment 简短的声音（节目）
6. the combination of text (the printed instruction), static graphic images (the color map), and digital sound (the narration describing traffic conditions) 包含了文本（打印指令），静态图像（彩色图）和数字语言（说明交通情况的叙述）
7. computer-based training (CBT) 基于计算机的培训
8. There is a new type of software category—edutainment. 现在有一种寓教于乐的新型软件。
9. business presentations 商业简报

3. Internet
（英特网）

A：Are you a netizen?

B：What do you mean by netizen?

A：It is a compound word composed of *net* and *citizen*. It is the result of the development of the Internet technology. A netizen is a person who takes great interest in the Internet and makes full use of it.

B：In this sense, I am not up to a netizen, because I don't often get on-line although I am interested in it. Besides netizen, what words have emerged along with the Internet?

A：E-mail, e-commerce, cyberspace, chatroom, website, dotcom and so on.

B：Would you like to tell me something about the Internet?

A：We have discussed some mass media such as radio and TV, which are basically "one-way" communications, while the Internet is inherently a "two-way" medium.

B：Why?

A：The ease with which a computer user can communicate with the rest of the world has created a sense of "global community", which is found almost nowhere else. People around the world post websites, exchange e-mails, and participate in online chats.

B：The world appears as a village because of the Internet.

A：You are right. Nowadays people get an Internet-is-the-world mood.

B：The Internet is used for e-mail and online chats?

A：It goes beyond that. In fact, the access to the Internet feels more and more like the utility of sorts like the phone line, the electricity, and the cooking gas. You pay for it as the daily necessities. These days, of course, no more self-respecting people would live without accessing to the Internet. The Internet access gives us our e-mail address, our capability of buying a book or a vacation online, even though the more traditional distribution channels are never less convenient and always more reliable. The Internet undoubtedly enriches our lives in many a way. The instant reach of the e-mail makes keeping-in-touch so rewarding and so excuse-proof.

B：You mentioned "dotcom" just now. What does it mean?

A："Dotcom" is used to describe a company which is related to the Internet. The 1990s was the golden age for Internet for it was heralded as another new earth-shaking technological revolution that would lead us to a very different kind of advanced world. Any dotcom company was warmly welcomed. The American stock market took off and went up.

B：I have seen two interesting cartoon pictures which may describe the situation you mentioned.

Two beggars are begging in the street side by side. One of them is hanging a sign with a word "beggar", while the other hanging a sign with "dotcom". The latter obviously gets more money.

A: It is the "dotcom" that makes difference. It seems as if any dotcom company would automatically become the company of the future equivalent of today's Microsoft. Any traditional company and business would be sneered at as passé. But the dream didn't last long and the Internet bubble finally burst and as a result, the American stock market went down, and many people lost much money.

B: But we should take advantage of the new technology to make our life more convenient.

A: Yes. Many companies have set up their own websites and put their information on-line so that it can be available to most people. For example, I can buy jeans directly from the factory without leaving home. First I go to a scanning studio where computerized 3D images of my body and my measurements are stored in a computer database. Then I get on line, open up the Just Jeans web pages, place orders, which will be delivered to my home.

B: Our life is greatly changed by the Internet.

A: Yes. Traditional business tends to find a way by combining with the Internet. For example, AOL "bought" Time-Warner. I put quotes around "bought" because I'd rather think AOL was smart enough to realize it couldn't survive by being AOL alone. It was ingenious for AOL to hitch onto the traditional entertainment giant Time-Warner and become part of the "new" old business, establish and secure a future for itself.

B: What is its future like?

A: In the next few years, you can expect a more dynamic Internet experience, faster connection speeds and freedom from a few of the hassles that accompany existing technology.

B: Can you name some of them?

A: Here are a few of the highlights: broadband connections, which increase the speed at which data is transmitted, should multiply quickly. VoIP technology, short form for Video-over-IP, which moves analog voice traffic into the Internet, will explode in popularity because of its great cost savings over traditional long-distance phone calls. Unified messaging—which ties your voice mails, faxes, e-mails and pager messages together so that you can retrieve any type of the message through any of the channels—will become more popular because of its potential time savings.

Notes

1. netizen 网民
2. compound 复合的
3. In this sense, I am not up to a netizen. 在这个意义上，我算不上一个网民。
4. emerged 出现

5. e-mail, e-commerce, cyberspace, chatroom, website, dotcom 电子邮件、电子商务、互联网空间、聊天室、网站、公司域名

6. The Internet is inherently a "two-way" medium. 互联网就是一种内在的"双向"媒体。

7. The ease with which a computer user can communicate with the rest of the world has created a sense of "global community" which is found almost nowhere else. 计算机用户轻松地与外界交流,产生一种"地球社区"的感觉,这种感觉在其他地方找不到。

8. online chats 网上聊天

9. an Internet-is-the-world mood 网络就是世界的情结

10. In fact, the access to the Internet feels more and more like the utility of sorts like the phone line, the electricity, and the cooking gas. 实际上,使用互联网越来越感觉像使用电话线、电、做饭用的天然气这样的日常用品。

11. These days, of course, no more self-respecting people would live without accessing to the Internet. 当然,现在任何自重的人如果离开互联网都难以生存。

12. The Internet undoubtedly enriches our lives in many a way. 毫无疑问,互联网在很多方面丰富了我们的生活。

13. The instant reach of the e-mail makes keeping-in-touch so rewarding and so excuse-proof. 电子邮件的及时传递使得保持联系变得那么有益,因此就没有不保持联络的托辞了。

14. It was heralded as another earth-shaking new technological revolution. 它预示着另一次震撼世界的新技术革命的到来。

15. stock market takes off 股市飙升

16. cartoon pictures 卡通漫画

17. It seems as if any dotcom company would automatically become the company of the future equivalent of today's Microsoft. 好像任何一个网络公司都会在将来自动地变成今天的微软公司。

18. Any traditional company and business would be sneered at as passé. 任何一个传统的公司或行业都可能被嘲笑为落伍。

19. bubble 泡沫

20. a scanning studio 扫描室

21. computerized 3D images 经计算机处理的三维图像

22. AOL was smart enough to realize it couldn't survive by being AOL alone. 美国在线很聪明地意识到只靠自己是难以生存下去的。

23. It was ingenious for AOL to hitch onto the traditional entertainment giant Time-Warner and become part of the "new" old business, establish and secure a future for itself. 美国在线机灵地傍上传统娱乐巨头时代华纳,变成这个"新"的旧行业的一部分,建立和保证了自己的未来。

24. In the next few years, you can expect a more dynamic Internet experience, faster connection speeds and freedom from a few of the hassles that accompany existing technology. 未来几年,你有希望得到更有活力的因特网体验,更快捷的连接速度,也不会出现现有的技术带来

的烦恼。
25. Here are a few of the highlights. 这是几个最突出的部分。
26. broadband connections 宽带接入
27. VoIP technology, short form for Video-over-IP, which moves analog voice traffic into the Internet, will explode in popularity because of its great cost savings over traditional long-distance phone calls. IP视频技术（Video-over-IP的缩写形式）将模拟声音输入因特网，比起传统的长途电话，它大大地节省了钱，将会大受欢迎。
28. Unified messaging—which ties your voice mails, faxes, e-mails and pager messages together so that you can retrieve any type of the messages through any of the channels—will become more popular because of its potential time savings. 统一信息传递技术将你的声音、传真、电子邮件和传呼信息结合在一起，你可以通过任何渠道重新收到任何种类的信息，由于节省时间的潜力，它将更受欢迎。

4. Internet-based Training
（基于英特网的培训）

A: What are you reading?

B: I'm reading something about the Internet. By the way, what is CBT?

A: Well, it stands for computer-based training, which allows employees to learn at their own pace and with little disruption of the workflow in the office.

B: Is it very popular?

A: It has been used by some companies for years, but CBT programs have been very costly. Now faced with fast-changing technology, a tight labor supply, and ever-increasing pressures on budgets, IT managers are looking to Internet-based Training (IBT) as a viable option for keeping employees' skills up to date.

B: Is that different from CBT?

A: Yes, Internet-based training (also called web-based or on-line training) has emerged as a less expensive and time-saving way to satisfy many companies' training needs.

B: Will it displace classroom training and CBT?

A: At least, it provides another option for IT departments to bring trainings to their employees.

B: Can you tell me more about it?

A: OK. These programs offer live, on-line classes either with an instructor and other students, or as an independent study course in which students work at their own pace. Internet-based training can not only deliver course materials similar to CBT, but also tends to use the medium's capability to offer interactivity. Most course programs are using Java, ActiveX, or proprietary browser plug-ins to re-create the classroom experience.

B: So in these live courses, students and an instructor can meet in the on-line classroom for instruction.

A: Right. They have several accesses to the online classroom. Course materials can be downloaded from some companies' online library or received via E-mail. Students are required to do assignments, which will be sent to their instructors through E-mail for evaluation and comment.

B: It seems to be a cost-effective and flexible way of taking courses.

A: Yes. Because most training is browser-based, it is available from most platforms.

Notes

1. Internet-based Training 基于因特网的培训
2. CBT: computer-based training 利用计算机的培训
3. at their own pace 以他们自己的速度
4. disruption of the workflow 干扰工作流程
5. tight labor supply 劳动力紧张、短缺
6. ever-increasing pressures on budgets 日益增加的预算压力
7. as a viable option for keeping employees' skills up to date 作为一种使雇员的技能赶上最新时代的可行选择
8. IT department 信息部门
9. live, online classes 实况在线课堂
10. to use the medium's capability to offer interactivity 利用这种媒体能力，提供交互式学习活动
11. Java, ActiveX, or proprietary browser plug-ins Java 技术、ActiveX 或者专用浏览器插入程序
12. download 下载
13. online library 在线图书馆
14. browser-based 基于浏览器的
15. platform 平台

5. Hot Technologies of Computer
（计算机热门技术）

A: I bought a book today.

B: What kind of book?

A: About the hot technologies of computer.

B: Oh, it must be very interesting. What does it say?

A: It introduces five hot technologies of computer. They are multimedia technology, parallel processing technology, object orientation programming technology, graphical user interface and artificial intelligence.

B: I think I know a little about multimedia technology. It is a technology that combines text, graphics, animation, audio and video on personal computers. But what is parallel processing technology?

A: It is a technology which tackles a task with hundreds or even thousands of microprocessors. Experts predict that it will make a small but forceful push into the commercial market in this decade. Computers will continue to get better at automatically "parallelizing" code and much effort will go into developing software tools for porting, developing and debugging software for parallel machines.

B: Now what is object orientation programming technology?

A: OK, let me open the book at page 32. Here it is. It refers to object-oriented programming systems, which allow programmers to assemble applications in modular. It can simplify systems development and maintenance because code can be reused for a variety of applications. More importantly, object-oriented technologies let businesses adapt quickly to critical business changes.

B: So they find appreciation in this decade. The book also tells what is graphical user interface, I suppose.

A: Yes, surely. It refers to voice recognition and speech synthesis, as well as pen-based computing. Many analysts agree that graphical user interface (GUI), especially Microsoft's Windows 3.0, will enjoy tremendous growth.

B: So it's possible to smooth the human-to-computer interface. How about artificial intelligence? Can you show me what the book says?

A: Certainly. Now read here. It refers to the intelligence in a machine, which is the ability to solve complex problems swiftly. The problems may cover medical diagnosis and prescription, resolving fiscal or legal matters, etc. As companies struggle for competitive advantages in an increasingly service-oriented market, artificial intelligence will continue to join the mainstream.

B: Well, I really enjoy talking with you very much today. But I think I've just learned a little about these technologies. Maybe I will borrow the book from you for more detailed information someday.

A: You're welcome.

Notes

1. hot technologies of computer 计算机的热门技术
2. multimedia technology, parallel processing technology, object orientation programing technology, graphical user interface and artificial intelligence 多媒体技术、并行处理技术、

面向对象的编程技术、图形用户接口和人工智能

3. text, graphics, animation, audio and video 文本、图形、动画、音频和视频
4. compilers 编译人员，编辑人员
5. to get better at automatically "parallelizing" code 使自动"并行化"代码方面变得更好
6. porting, developing and debugging software for parallel machine 并行机的软件移植、开发和调试
7. to assemble applications in modular 以模块方式汇编应用程序
8. find appreciation 增值，获得赞赏
9. voice recognition and speech synthesis, as well as pen-based computing 声音识别、语音合成以及用笔输入的计算
10. to smooth the human-to-computer interface 实现通畅的人机界面
11. medical diagnosis and prescription, resolving fiscal or legal matters 医疗诊断和处方，解决财政和法律事务
12. struggle for competitive advantage in an increasingly service-oriented market 在日益以服务为导向的市场里争夺竞争优势
13. to join the mainstream 加入主流

6. E-life
（电子化生活）

A: Do you know anything about e-commerce?
B: Oh, yes, I know a little about it. But now there're many e-things such as e-business, e-book, e-shopping, e-mail, e-market etc. The Chinese public is rapidly turning its attention to on-line shopping.
C: Now everything can be done via computer by Internet or network. Shoppers have adapted their walk-and-see shopping pattern to searching for goods at various new sites on line.
B: Yes, according to a special report in *American Newsweek* in October Ⅱ issue, 1999, it summarizes all e-things in one word, that is *E-life*. I've read this report, and I think the word *E-life* is exactly right. The title of this report is "the Dawn of *E-life*".
A: Say something more about it.
B: Ok, the report says *E-life* is coming or some of *E-life* has already come. The dawn of *E-life* isn't just about the future—it's about the here and now. The Internet has started transforming our way of life. We may stay at home for schooling, for seeing a doctor, for shopping and for reading all kinds of newspapers and magazines and so on.
A: Do you know which corporation is the first for Internet commerce in the world?
C: It's called "Amazon.com". It's called the flagship for e-commerce. Do you know the person

who first created the on-line shopping?

A: I don't know.

C: Jeff Bezos, he is the founder of Amazon.com, and started on-line shopping in July, 1995. He started his on-line bookstore in a garage. Bezos is a person who not only changed the way we do things but helped pave the way for the future. The *New York Times* announced him "Man of the Year" in 1999. Bezos is at the forefront of on-line retailing.

A: Could you explain e-commerce with simple words?

B: E-commerce may be simply something about selling goods directly to consumers on the net and buying goods directly from sellers on the net. That's to say you needn't go to the shops or stores. What you want will be delivered to your house. You may get a cup of coffee delivered from the corner store downstairs by on-line shopping.

C: That's about e-shopping. E-shopping (originally called tele-shopping) is shopping without leaving home, placing orders by electronic means. Customers have access to some kind of electronic catalogue that they can bring up on a screen via the telephone network. They then place orders via keyboard or key-pad which are transmitted automatically to the suppliers' ware-house. Goods are then delivered by van or mail.

A: What else can be done?

C: Besides, you may have on-line dining reservations, ticket reservations and even online donations to "Project Hope" and in the U.S., voters may stay at home to elect their president.

A: What about e-commerce in our country?

B: We actually have already started this business. China has made a package of measures and also a government on-line project officially to highlight China's determination to promote e-commerce. The Ministry of Information Industry will work to develop and introduce equipment and software to create a sound technical environment for the development of e-commerce.

C: And since the second half of 1999, the popularity of 8848.net has become an example and incentive for other Internet companies to join in e-commerce business. By the end of 1999, about 571 Chinese websites, about web-based stores in our country had been involved in electronic commerce.

A: I think it's imperative for China to use e-commerce to remain competitive in the world. Shopping on-line is still not easy enough for customers. Some of web surfers who want to buy something on-line don't know how to place an order. Because buying on-line is still too complicated. So keeping e-commerce simple in operation is very important.

C: I think when we talk about e-commerce or e-life, we mustn't forget e-mail. When online users are asked what they do on the net, e-mail is always No. 1.

B: Quite right. If we accept that the creation of the globe-spanning Internet is one of the most important technological innovations of the last half of the 20th century, then we must give e-mail pride of place.

C: E-mail is both the catalyst and the instrument of the change.
A: Who invented e-mail, do you know that?
B: E-mail is invented and first used by an American named Ray Tomlinson. He was first employed to work for Arpanet in 1968 which was the predecessor of Internet. It is Tomlinson who first used the symbol @ and sends an e-mail to himself.
C: I also know something about Tomlinson. He graduated from MIT in 1965 and then spent 2 years to obtain his doctorate degree in computer engineering and then he worked in an enterprise affiliated to American army, engaging in computer research. He is still living.
A: Oh, thank you very much. I think I've learned a lot from you today.

Notes

1. in *America Newsweek* in October Ⅱ issue, 1999 于1999年10月第二期在《美国新闻周刊》发行的
2. It summarizes all e-things in one word, that is *E-life*. 它将所有电子事物概括为一个词——电子化生活。
3. "the Dawn of *E-life*" 电子化生活的黎明
4. It's about the here and now. 就在此时此地。
5. Amazon.com 亚马逊公司网站（总部设在美国西雅图、华盛顿，亚马逊公司是提供网络购物服务的第一家大公司）
6. the flagship for e-commerce 电子商务的旗舰
7. Jeff Bezos 世界著名电子商务公司亚马逊的执行总裁
8. The forefront of on-line retailing 在线零售的最前列
9. Goods are then delivered by van or mail. 商品由送货车或邮件方式交货。
10. on-line dining reservations 在线就餐预订
11. on-line donations to "Project Hope" 在线资助"希望工程"
12. a package of measures 一整套措施
13. a government on-line project 一项政府在线工程
14. the popularity of 8848.net 受欢迎的8848网
15. some of web surfers 一些上网者
16. the creation of the globe-spanning Internet 创造覆盖全球的互联网
17. pride of place 最重要的地位
18. catalyst 催化剂，刺激因素
19. Ray Tomlinson 托姆林森（发明e-mail的美国人，是第一个使用@符号发电子邮件的人）
20. Arpanet in 1968 which was the predecessor of Internet 1968年创建的Arpanet是互联网的前身
21. MIT (Massachusetts Institute of Technology) 麻省理工学院
22. an enterprise affiliated to American army 美军所属企业

7. E-commerce
（电子商务）

A senior engineer (E) of On-line Development is talking about E-commerce with the managers of marketing (M1, M2...).

E: E-commerce is heard frequently in modern society. It is one of the most common business terms in use as we embark for the 21st century. E-commerce and globalization are two dynamic forces shaping our world today. One of their greatest impacts is in interweaving many countries' markets together. The online technology has changed business infrastructure. Corporations have to learn to market themselves in other languages in order to be on the competitive edge. The Internet is one of the various marketing tools that can be used as a medium to extend business transactions and create new opportunities. E-commerce has become a virtual reality in marketing products and services over the globe. In terms of some recent survey, by 2005, there will be around one billion people online worldwide.

M1: Exactly, if we don't jump on the bandwagon now, we will miss the boat.

M2: Ms. M1, can you give us an estimated figure about e-commerce spending in the upcoming years?

M1: Ok, some experts predict that by 2005, about three quarters of e-business income will generate outside the U.S.A. With this in mind, it's an absolute prerequisite to translate and promote our website in the language of the target country.

M2: What do you mean by doing that?

M1: It is true that many people abroad read and speak English. But whether a person speaks English or not has nothing to do with the responsibility of a website to communicate in the language of the target markets, simply because these non-English natives still prefer to surf online in their own language.

M2: In other words, they live their life in their own language, not in English.

E: Yes, if you want to market to them, you need to utilize online marketing techniques to get their attention. The first step is to register your translated web pages in the country's specific indices, search engines, and web directories.

M1: Mr. E, what are some of the most cost-effective marketing tools you can apply to on the web?

E: Very good question! The best way to get free publicity is through the press. The news media and general public are often keen on the latest Internet news.

M2: Many companies also use affiliate programs and banner advertising to increase e-commerce sales. Newsgroups and public forums can also generate considerable interests in your target countries.

E: Some of the other common and effective online tools are: zines, e-flyers, collaborative marketing, hosting services, virtual malls, and conferences.

M1: The intranet in particular is one of our biggest cost savings e-business feature. Its return on investment is 32% since we've integrated the system.

E: Yes, the increase in efficiency and consistency of information was due mainly in part to production, inventory control, shipping and receiving, and design.

M2: We have seen the same result with our integration of the extranet. Since then, our business-to-business operations have taken off.

E: That's it. This is an area we can really benefit as we expand our operation to the foreign market. The automation of extranet with outside partners, customers, distributors, and suppliers will be a major factor to our success abroad.

M1: It is important to remember that when we move into a foreign market, we must be careful to comply with all applicable export and import requirements as well as an awareness of cyberlaw issues.

M2: What do cybersellers pay special attention to?

M1: When consumers go online, they need to trust the vender. Although we already have consumer confidence in our product, buying online is different from buying in a store. Cybersellers have to develop an environment of trust. While we ask people to type in their credit card number, they have no idea who'll have that information. So we must find some way to assure them that the data is secure.

M2: What is necessary to do on-line transactions?

E: To do on-line transactions, you'll need to link up with a financial institution, such as a bank or investment bank. They're the intermediaries between you and your customers' money. We'll use the same one that issues our store credit cards, therefore this will help with the trust factor.

M1: What kind of financial software should the customer buy?

E: He will need off-the-shelf financial software. If he has customized financial software, it may not have interoperability, because the financial data has to be shared with his financial institution.

M2: And what about encryption methods?

E: Generally most companies are going with one or two types of encryption. If necessary, the customer should check into off-the-shelf financial software.

M1: In speaking of foreign markets, Mr. M2, can you give us an assessment of e-commerce environment in China?

M2: Though China remains a developing country, the ambitious use of high technology has made in-roads with the growth of governmental and business-to-business forms of e-commerce. Government of all levels seek to use technologies to inform the public about laws, deal with customs and simplify procedures. In respect to business, direct marketing and sales online have begun despite the lack of credit card usage and distribution difficulties. Regardless of the nascent stage of e-business, China offers a huge potential market for companies using or selling related information technology products.

E: The importance of marketing our website to promote the digital gadget cannot be overemphasized. It is advisable to make certain that we continue to monitor the international index sites where our URLs are listed and keep at online promotion work in the countries that we are targeting.

M1: I think it is critical to establish a monthly budget for our international website promotion, as more visitors turn into sales. E-business is no longer considered as an alternative but essential for business survival.

Notes

1. as we embark on the 21st century 当我们进入 21 世纪
2. two dynamic forces shaping our world today 改变当今世界的两大动力
3. interweaving many countries' markets together 将许多国家的市场紧密结合在一起
4. business infrastructure 工商企业的基础设施
5. be on the competitive edge 参与竞争
6. jump on the bandwagon 顺应潮流
7. surf online 在线漫游、浏览
8. register your translated web pages in the country's specific indices, search engines, and web directories 将翻译的网页在目的国的索引、搜索引擎和网络导航上登录
9. the most cost-effective marketing tools 成本最低的市场营销手段
10. affiliate program 联营项目附属机构计划
11. banner advertising 旗帜广告
12. newsgroups and public forums 新闻集团和公众论坛
13. zines (online magazines) 在线杂志
14. e-flyers 电子广告传单
15. hosting services 网络服务公司提供的业务广告特殊服务
16. virtual malls 虚拟购物中心
17. intranet 内部网
18. the increase in efficiency and consistency of information 信息的效能和一致性的提高
19. extranet 外部网
20. take off 开始受欢迎，开始有名了
21. an awareness of cyberlaw issues 网络法方面的意识
22. cyberseller 互联网销售商
23. to trust the vender 信任卖主

24. an environment of trust 信任环境
25. off-the-shelf financial software 现成的财务（金融）软件
26. customized financial software 定做的财务软件
27. interoperability 互相间的可操作性，双方共用性
28. encryption methods 加密方法
29. check into off-the-shelf financial software 检索到现成的财务软件中
30. in-roads 进展
31. nascent stage 初期阶段
32. the digital gadget 数字装置
33. URLs（An acronym for universal resource locator）通用资源定位器的缩写

8. Office Automation
（办公室自动化）

A: Do you know anything about office automation?

B: Yes. Office automation is the application of computer and communication technology to improve the productivity of clerical and managerial office workers.

A: What are the major functions of an office automation system?

B: Well, it includes text processing, electronic mail, information storage and retrieval, personal assistance features, and task management.

A: By what means does an organization realize its automation?

B: Today's organizations have a wide variety of office automation hardware and software components at their disposal. The list includes telephone and computer systems, electronic mail, word processing, desktop publishing, database management systems, two-way cable TV, office-to-office satellite broadcasting, on-line database services, and voice recognition and synthesis. Each of these components is intended to automate a task or function that is presently performed manually.

A: Is there any criteria for office automation?

B: Experts agree that the key to attaining office automation lies in integration—incorporating all the components into a whole system so that information can be processed and communicated with maximum technical assistance and minimum human intervention. This goal can be accomplished when computer, communication, and office equipment are networked and an office worker can easily access the entire system through a personal computer sitting on his or her desk.

A: How about the most popular applications of office automation?

B: They are desktop publishing, video conferencing, and videotex. Now the facilities of the electronic office have been used to increase the efficiency of external communications, such as telex, external e-mail, facsimile transmission, electronic data interchange, and view data.

A: I'm convinced that office automation will change substantially the way people work in an office.

B: Yes, I quite agree.

Notes

1. Office automation is the application of computer and communication technology to improve the productivity of clerical and managerial office workers. 办公室自动化是应用计算机和通信技术提高办事人员和管理人员工作效率的技术。
2. text processing, electronic mail, information storage and retrieval, personal assistance features and task management 文本处理、电子邮件、信息存储和检索、个人事务辅助管理和任务管理
3. at their disposal 任他们处理
4. word processing, desktop publishing, database management systems, two-way cable TV, office-to-office satellite broadcasting, on-line database services, and voice recognition and synthesis 文字处理、桌面印刷系统、数据库管理、双向电缆电视、办公室对办公室的卫星广播、联机数据库服务、声音识别及合成系统
5. criteria 标准
6. integration 综合性
7. Incorporating all the components into a whole system so that information can be processed and communicated with maximum technical assistance and minimum human intervention. 将各硬、软件紧密结合成一个完整的系统，使得信息处理和通信获得最大的技术支援，人为干预最少。
8. video conferencing, videotex 电视会议、信息传视
9. telex, external e-mail, facsimile transmission, electronic data interchange, and view data 电传、外部电子邮件、传真发送、电子数据交换、可视数据

9. Global Positioning System (GPS)
(全球定位系统)

A: Recently I read an article in a newspaper. In it there's a term GPS. Do you know what is GPS?

B: Yes, I know a little. It's the new Global Positioning System for locating things. GPS is the short form. It functions through the use of 24 satellites that are circling the earth. The satellites transmit their location and time 1 000 times a second. A computer on earth receives this information from several satellites and with that information precisely calculates your latitude, longitude and altitude.

A: How precise?

B: At first it could be accurate to within 100 meters, but now it can pinpoint the location to within a few feet. Some cars was equipped with a computerized map linked to a GPS receiver hidden in the car. If you want to know how to get to your destination, the system will show the fastest route. A yellow arrow appears on a screen and an electronic voice will give directions.

A: If I get lost in a certain place, such as a huge grassland or desert or have a breakdown with my car in a remote area, can it help?

B: Sure, it is widely used. Some cars are now equipped with "Mayday" buttons. When pushed, the car's cellular phone automatically transmits the car's identification number and location to secure people who then call the police, ambulance or whatever is needed. It was first used by a ford engineer whose test car died in a parking lot. His call for help was received by security personnel in Texas who called a tow truck in Michigan and within minutes a truck was there to help.

A: Well, it's really very useful. Can it be used in military affairs?

B: Yes. GPS has been used successfully in wartime situations too, like the war of America against Afghan terrorists and the War on Iraq from March to April, 2003.

A: Besides military uses and uses in cars, is it used in peace time?

B: Yes. It has many uses in peace time which are the most important. It is said to have been developed to give GPS eyes to a blind. They combine a GPS receiver and laptop computer with a speech synthesizer which are put in a backpack and connected to the person with headphones. An electronic voice will give the person directions on where to go. It is like a sound guide to the blind person. I've heard of other wonderful uses for industry, farming and even ocean wreckage retrieval.

A: Oh, I see. High technology is making great changes on the way we live. I've learned a lot from you today, thank you very much.

Notes

1. global positioning system (GPS) 全球定位系统
2. locating 找出；确定位置
3. latitude, longitude and altitude 纬度、经度和海拔高度
4. pinpoint the location 精确定位
5. a GPS receiver 全球定位接收器
6. have a breakdown with my car in a remote area 我的车在偏远地区突然出现了故障
7. "Mayday" button 无线电求救信号键
8. cellular phone 移动电话
9. ambulance 救护车

10. a tow truck 拖车
11. laptop computer 小型计算机
12. a speech synthesizer 语言综合处理器
13. backpack 背包
14. headphones 耳机，头戴式受话器
15. ocean wreckage retrieval 海难营救，海难搜索

10. Digital Technology
（数字技术）

A: (Reading a Chinese newspaper)

B: What are you reading? You're so absorbed in your reading.

A: Have you heard of the concept "digital earth"? A symposium on digital earth is being held in our Beijing.

B: What's digital earth?

A: Have you read Gore's speech?

B: I don't know the speech at all and who is Gore.

A: That's about digital earth, which is a new concept in information technology. Gore is the vice president of United States. His full name is Al Gore.

B: Oh, that's interesting, digital earth, what sort of thing is it?

A: It is very difficult to give a clear definition. The newspaper also has not given us a clear definition. It is considered to be a strategic thought based on information technology such as information superhighway, etc. American vice-president Al Gore first suggested this concept in the speech made on January 31, 1998 in California Science Center.

B: It seems this term is very funny. I think we can say the digitalization of our globe, or everything goes digital. But how about digital earth?

A: I think digital earth is an abstract concept. According to Gore's speech, it is a multi-resolution, three-dimensional representation of the planet, into which we can embed vast quantities of geo-referenced data.

B: Is that only an imagination?

A: Eh, yes, my understanding of why we call a digital earth is that it's actually a frame by using three-dimension version of our earth. That's to say we consider our earth as a kind of frame and we can search for the information we want from any part of this frame through high-speed networks.

B: That's good. But is it realized in United States?

A: Not yet. Mr. Al Gore said in his speech that digital earth would not happen overnight. We

should do a lot of things.
B: A lot of things? What are they?
A: First, we should focus on integrating the data from multiple sources that we already have; second, we should endeavor to develop a digital map of the world at 1-meter resolution. In this way we could put the full range of data about our planet and our history at our fingertips.
B: And we could conveniently get the information we want quickly. When could we do in this way?
A: I don't know. But the digital technology is developing rapidly and now we have digital television, digital camera and different kinds of digital devices. So I think the time for it will come soon.
C: Digital technology is actually a study and development of devices that store and manipulate numbers. It's different from analog(ue) technology which involves devices that function without storing or using numbers. But digital devices store numbers using only two digits, 1 and 0, called binary numbers. Every digit in a binary number represents a power of two.
A: Now digital experts have developed a new digital infrastructure. It's like the human nervous system. Moving digits will put you on the leading edge of a shock wave of change that will shatter the old way of doing business.
C: It's said, "A digital nervous system will let you do business at the speed of thought — the key to success in the twenty-first century".
B: How and where do you know this new concept?
C: I read from a book *Business @ the speed of thought* written by Bill Gates.
B: Can you lend me the book?
A: And me?
C: Oh, I can't. Actually I read this book from the Internet. But I have downloaded its outline and the summary of the book. I can print it out and lend it to you two.
B: Thank you very much, and it's kind of you to print it for us.
A: Thanks a lot.

Notes

1. the concept "digital earth" "数字地球"的概念
2. Gore's speech 戈尔的讲话（Gore 的全名是 Al Gore，原美国副总统）
3. the digitalization of our globe, or everything goes digital 全球数字化或一切都数字化
4. an abstract concept 抽象概念
5. It is a multi-resolution, three-dimensional representation of the planet, into which we can embed vast quantities of geo-referenced data. 这是一个对地球具有高分辨率、三维显示的框架，我们可以嵌入（输入）供全球参考的大量数据。
6. a frame by using three-dimension version of our earth 一个用三维方式描述地球的框架

7. put the full range of data about our planet 输入有关我们星球完整的数据
8. fingertips 手指尖
9. manipulate numbers 控制数字
10. It's different from analog(ue) technology which involves devices that function without storing or using numbers. 它是与没有储存或使用数字功能装置的模拟技术不同的。
11. Every digit in a binary number represents a power of two. 二进位数的每个数字表示许多2（或2的乘方）。
12. Moving digits will put you on the leading edge of a shock wave of change that will shatter the old way of doing business. 移动的数字将把你推到出奇制胜的变化震激波的前缘，将使你做生意的老办法破灭。
13. a digital nervous system 数字神经系统
14. *Business @ the speed of thought* written by Bill Gates 比尔·盖茨写的一本书——《思维速度的工商企业（商贸业务）》

11. Automation & IT Application
（自动化和信息技术的应用）

A: No doubt, automation & IT is having a great impact on all types of industries.

B: Yes, some people see it as the second revolution in the industrial process. The first was the assemble line system.

A: That's an interesting view. And actually automation & IT has been widely used not only in the manufacturing processes but also in many other areas of business distribution, warehousing, material handling and physical control etc.

B: Very true. I think it will be applied to broader areas, for it not only saves labor, but improves efficiency and output.

A: That's true. The movement towards automated & IT process must carefully take many things into consideration.

B: Right, especially the capital expenditures are involved. Automation & IT must yield positive results and those results must justify the costs.

A: Yes, this is true especially in industrial manufacturing applications where the installation of automated & IT process is usually extremely expensive.

B: That's correct. In such applications, automation & IT must frequently be seen as an integrated process. In other words, automation & IT in only one sphere of the manufacturing process can sometimes lead to real production bottlenecks.

A: I don't follow you here. Why would that be so?

B: Well, automation & IT will presumably increase the output, but finished product output will depend on material input and the ultimate distribution of the product output to the next manufacturing stage.

A: Oh, so input in distribution could be a bottleneck.

B: Right. That's why careful integration is necessary.

A: Yes, but that's the function of the feedback equipment in designing self-regulating controls to operate the equipment. In your drill press, for example, feedback from the machine will control devices to either slow down output or speed up input.

B: Yes, that's true, only if the output distribution and material input processes are automated. If the input and output processes are not automated, however, additional labor will be necessary or the automated processes will have to be slowed down and this will lead to a reduction in productivity.

A: I see what you mean. This integration that you talked about must involve both the related automated processes and the non-automated processes.

B: Exactly. That's why the feedback control mechanisms are so important. And all parts of the system must be responsive to the feedback.

A: And that's what you mean by careful analysis of the entire process when automated & IT equipment is installed.

B: Yes, that's where the essential elements of integration our plant mapped out.

Notes

1. automation & IT 自动化和信息技术
2. impact 影响
3. assemble line system 装配线系统
4. warehousing 仓储
5. physical control 实物控制（管理）
6. The movement towards automated & IT process must carefully take many things into consideration. 向自动化和信息技术过程迈进时，必须细致考虑众多的事情。
7. yield positive results 产生积极效果
8. justify 证明……是合理的
9. industrial manufacturing application 工业制造业的应用
10. integrated process 一体化的程序
11. sphere 方面；范围
12. Automation & IT will presumably increase the output, but finished product output will depend on material input and the ultimate distribution of the product output to the next manufacturing stages. 自动化和信息技术可能会提高钻床产量，但是成品产量将取决于原材料的投入和最终流转到下一个生产阶段的产品产量。
13. feedback equipment 反馈设备

14. self-regulating control 自动调节控制
15. Feedback from the machine will control devices to either slow down output or speed up input. 机器的反馈系统将促使控制装置放慢输出或加速输入。
16. output distribution 产量分配
17. reduction in productivity 劳动生产率的下降
18. the feedback control mechanisms 反馈控制机制
19. responsive to the feedback 响应反馈
20. That's where the essential elements of integration our plant mapped out. 这正是我们工厂制定有关一体化的基本要素。

12. Computer Simulation
（计算机模拟）

S — a student　　　　P — a professor

S：Would you please explain what computer simulation is?
P：Well, computer simulation is based on mathematical (math) models representing the nature of the object under study or examination. The math model comprises a series of equations that depict the inherent process of the object in math terms. A computer simulation program includes algorithms that are derived from those equations. The outcome of simulations is usually expressed in rather abstract forms, for example, 2-D diagrams, curves, tables, figures, and they are difficult for those who are unfamiliar with math and abstract concepts to understand.
S：What are the advantages of computer simulation in technology and engineering?
P：This is a valuable question. Computer simulation, as a powerful analytical tool is widely used in scientific research and engineering design, and demonstrated unrivalled advantages. With computer simulation, scientists and engineers do not have to build real hardware models or primary prototypes when they observe an unknown phenomenon, analyze a complex process, and design a machine or a building etc. Computer simulation is particularly significant when the object under study and examination is costly or even impossible to be built into a real model. For instance, to study the cause of engine malfunction that has led to a series of supersonic planes crashed, or to examine the impact on passengers when an airplane crashes, researchers may have to repeat simulating the calamities over and over again before they find out what a conclusion they need to reach. Apparently, these can only be simulated by running computer simulation programs or the like, rather than replicating the tragedies.
S：Any more examples?
P：Another example is engineering design, in which engineers have to try many schemes and

parameters before they come up with a satisfactory design. Using computer simulation programs, engineers can accomplish that iterative process each time by inputting different schemes and parameters into their computer models, rather than building many different real models.

S: How about the practical application of computer simulation to technology, engineering and design?

P: Here is an example. To analyze the distribution of stress in a fuselage when the plane is flying, the computer simulation package will first set up a mathematical model for this specific theme which comprises equations derived from aerodynamics, elasticity. Structural mechanics then implement a series of computations on a simplified structure of the plane, based on the finite element stress analysis. Finally, it gives the outcome which will be clusters of curves spread over the simplified structure of the plane, each indicating the locations in the fuselage that suffer stress of a uniform value. The accuracy of the simulation depends on the combinatorial accuracy of the mathematical model, that is, how closely the model is built to represent the real plane and its environment in terms of math, geometry and mechanics.

S: Have computer graphics techniques helped computer simulation by creating realistic 3-D images to depict the object to be simulated and the environment around it or the effect imposed on it over the past two decades?

P: Sophisticated computer simulation package capable of providing real time interactive images have emerged, although still beyond the reach of most industries. For the moment, in many cases, animated images for research and engineering design constitute simply an additional presentation facility within a 3-D modeling system. In others, moving images and 3-D images aid the understanding of computer simulation results. Meanwhile, many CAD (computer aided design) systems have incorporated visual modules to enable engineers to interactively "walk" through their 3-D pseudo models on screen to review their designs and present them to their clients. This convenience is particularly important to architects. In the design and planning of automated manufacturing systems, 3-D graphical simulation allows rapid kinematic and dynamic analyses of alternative mechanism designs, and offers a versatile tool for visualizing and assessing alternative system layouts. Most CAD packages for complex parts-machining have an animation capability.

Notes

1. computer simulation 计算机模拟技术
2. mathematical models 数学模型
3. comprises a series of equations that depict the inherent process of the object in math terms 包括用数学语言描述的物体固有（内在的）过程方程式
4. algorithms 演算法（计算过程）

5. derived from 从……得到
6. The outcome of simulations is usually expressed in rather abstract forms. 模拟结果通常用比较抽象的形式表示。
7. 2-D (2-dimensional) diagrams, curves, tables, figures 二维图形, 曲线, 表格和数字
8. difficult for those who are unfamiliar with mathematics and abstract concepts to understand 对那些不熟悉数学和抽象概念的人是难以理解的
9. unrivalled advantages 无与伦比的（极佳的）优点
10. real hardware models 仿真的硬件模型
11. primary prototypes 最初的原型机（样品）
12. engine malfunction 引擎故障
13. supersonic planes crashed 超音速飞机坠毁
14. the impact on passengers 对乘客的冲击
15. the calamities 灾难（空难）
16. replicating the tragedies 再现（复制）这场灾难（悲剧）
17. schemes and parameters 计划（方案）和参数
18. come up with 想出, 提出
19. iterative process 重复过程
20. to analyze the distribution of stress in a fuselage 要分析飞机机身应力的分布
21. the computer simulation package 计算机模拟软件包
22. this specific theme 这一特殊主题（课题）
23. comprises equations derived from aerodynamics, elasticity 包括从空气动力学, 弹性力学得到的方程式
24. the finite element stress analysis 有限元应力分析（工程设计中的结构分析法。它将一个复杂的工程结构——建筑、机械等, 分成有限个容易表述的子结构, 称为有限元, 将有限元及其相互间的关系用方程组的形式表示, 输入计算机程序, 计算应力的大小）
25. clusters of curves spread over 分布在……上的曲线束（簇）
26. suffer stress of a uniform value 承受相同值的应力
27. the combinatorial accuracy of the mathematical model 数学模型的组合精确度
28. geometry 几何学
29. computer graphics techniques 计算机图形技术
30. by creating realistic 3-D images to depict the object to be simulated and the environment around it or the effect imposed on it 用创造逼真的三维图像描述被模拟的物体及其周围的环境或对该物体施加影响的方法
31. sophisticated computer simulation package 先进的计算机模拟软件包
32. real time interactive images 实时交互（互动）图像
33. beyond the reach of most industries 尽管仍然超越了大多数应使用的能力
34. animated images 动画图像, 仿真图像
35. constitute simply an additional presentation facility within a 3-D modeling system 在三维模型

建造系统内自然地构成了另一种展示技巧

36. Many CAD (computer aided design) systems have incorporated visual modules to enable engineers to interactively "walk" through their 3-D pseudo models on screen to review their designs and present them to their clients. 许多计算机辅助设计系统已将可视模件合并，使工程师们能互动地"走"进屏幕上他们的三维虚拟模型，以评价他们的设计并将其展示给他们的客户。
37. architects 建筑师、设计师
38. automated manufacturing systems 自动制造系统
39. 3-D graphical simulation 三维图形模拟
40. kinematic and dynamic analyses of alternative mechanism designs 可供选择的机械设计的运动学和动力分析
41. offers a versatile tool for visualizing and assessing alternative system layouts 提供了一种具有多种技能的工具，使可供选择的系统外形设计具体化并能对其进行评价
42. Most CAD packages for complex parts-machining have an animation capability. 大多数用于复杂机器部件的计算机辅助设计软件包都具有仿真（动画）能力

13. Intelligent Machines
（智能机器）

S — a student   E — an engineer

S: Mr. E, having little knowledge about artificial-intelligence machines, I want to ask you a question.
E: Go ahead.
S: Can machines be as intelligent as human beings?
E: Recently, there is a great deal of discussion about smart machines and their relations to human intelligence. Some engineers and technicians are developing machines that can be programmed to think, to see, to learn and act in some ways as human beings. While working on their projects, inventors are learning more about a computer's abilities in relation to a human's. Some experts in MIT and University of Southern California are developing various smart machines. They will offer certain kinds of services and advice like a thoughtful assistant. An intelligent machine would be capable of helping people in a variety of professions. For instance, it can work tirelessly to help a doctor diagnose a person's illness. It might help a lawyer fight a court case and even make up a quiz for an English professor. In the future, those people who are psychologically disturbed might consult an artificial-intelligence psychologist for advice.
S: I read to know that when a team of scientists working in Antarctica wanted to explore the

volcano Mount Erebus, they designed a smart machine called Dante to do the dangerous jobs. Would you please describe Dante roughly for us?

E: Actually, Dante is a robot. It is more than 9 feet long, over 5 feet wide, and weighs almost 1000 pounds. It looks like a giant spider with 8 legs that can move in all directions at the speed of 6.5 feet a minute. In the center of its body is a pole holding 6 video cameras, three looking forward and three looking backward, giving a 360° view of an area.

S: How did the scientists dictate Dante to carry out the exploration of the volcano Mount Erebus?

E: They gave Dante instructions on how to climb down a steep slope, how to step over rocks and how to avoid dangerous objects in its path. They calculated that it would take Dante 24 hours to complete its journey 700 feet down. Upon arriving at the bottom of the volcano, Dante would take temperature readings, measure the composition of the gases, and collect samples of minerals. Filled with the knowledge of how to do everything, Dante seemed ready for its job. Placed at the edge of the volcano, it started down. But after travelling only a few feet, Dante stopped dead. The cable connecting it to its controllers had broken. Even the smartest, the most extraordinary robots, like people, can become disabled.

S: Did the scientists and Dante overcome the difficult problems?

E: No, not at that time. Dante's controllers were disappointed with their experiment, but they didn't give up. They said that they could learn from their failures, and as they got smarter, so could Dante. An improved Dante can then provides them with information never before available about the gases and minerals inside a volcano. The gases and minerals in Mount Erebus may be responsible for the hole in the ozone layer that develops over Antarctica each year. The hole allows increased rays from the sun that can cause cancer. Dante is an important new tool that can help solve some of our environmental problems. The scientists also say that someday more advanced descendants of Dante will be sent to explore the planet Mars.

S: What other jobs can smart machines do?

E: They may do ordinary jobs. They can be used to explore or exploit minerals on the ocean floor or in deep areas of mines too dangerous for humans to enter. They will work as firemen, housekeepers, and security personnel. They may be robot gas-station attendants that can fill a tank without the driver leaving the car. They can be programmed to manufacture various components and products, such as cars, planes, tractors, all kinds of machines etc. Smart robot dressmakers will conduct a dress to exact measurement while the woman is having a cup of coffee in the dress shop. Perhaps everyone will have a personal robot for the housework. Imagine what a robot will do for you!

S: Can science and technology take us too far by programming intelligence into machines?

E: Well, this is a major concern for many people. Some people think there is a danger in this. They say that these machines will take over their jobs and leave people unemployed. Scientists argue that artificial intelligence will free people from difficult, dangerous, or boring work. Then people can use their abilities and time for more creative occupations as well as doing exercises and participating in recreational activities. Some people worry about that intelligent machines or robots will become smarter than humans and turn into evil monsters. Scientists answer this fear simply: People will always be in control of "smart" machines. The prediction is for a good working relationship among human beings and their smart machine friends and tireless helpers.

Notes

1. intelligent machines 智能机器
2. artificial-intelligence machines 人工智能机器
3. like a thoughtful assistant 像一位有思维的助理
4. diagnose a person's illness 诊断病人的疾病
5. a lawyer fight a court case 律师在庭审中进行辩护
6. a quiz 小测验，问答比赛（游戏）
7. an artificial-intelligence psychologist 人工智能心理学家
8. in Antarctica 在南极洲
9. to explore the volcano Mount Erebus 考察埃里伯斯火山
10. a giant spider 一只巨大的蜘蛛
11. a pole holding 6 video cameras 装有（上面容纳）六个摄像机的杆
12. to climb down a steep slope 从陡坡往下爬
13. the composition of the gases 气体的组成成分
14. filled with the knowledge of ... 赋予机器人以……知识
15. Dante stopped dead. Dante 出现了致命的故障。
16. the smartest, most extraordinary robots 最具智能化、最特别的机器人
17. responsible for the hole in the ozone layer 对臭氧层洞的产生负有责任
18. more advanced descendants of Dante 更为先进的下一代 Dante
19. the planet Mars 火星
20. Robot dressmakers 机器人裁缝
21. evil monsters 恶魔

14. Robot and Its Recent Development
（机器人及其近期的发展）

S — a student   E — an engineer

S: I know little about robots. Could you give us a general introduction about robots?

E: No problem. Robots, becoming increasingly prevalent in industrial plants throughout the developed world, are programmed and engineered to perform industrial tasks without human intervention. Most of today's robots are employed in the automotive industry, where they are programmed to take over such jobs as welding and spray painting automobiles and truck bodies. They also load and unload hot, heavy metal forms used in machines casting automobile and truck frames. Robots, already taking over human tasks in the automobile field, are beginning to be seen, although to a lesser degree, in other industries as well. There they build electric motors, small appliances, pocket calculators, and even watches. The robots used in nuclear power plants handle the radioactive materials, preventing human personnel from being exposed to radiation. These robots are responsible for the reduction in job-related injuries in this new industry.

S: What makes a robot a robot and not just another kind of automatic machine?

E: Robots differ from automatic machines in that after completion of one specific task, they can be reprogrammed by a computer to do another one. For example, a robot doing spot welding one month can be reprogrammed and switched to spray painting the next. Automatic machines, however, are not capable of many different uses, they are built to perform only one task.

S: How about the next generation of robots?

E: They will be able to see objects, will have a sense of touch, and will make critical decisions. Engineers skilled in microelectronics and computer technology are developing artificial vision for robots. With the ability to "see", robots can identify and inspect one specific class of objects out of a stack of different kinds of materials. One robot vision system uses electronic digital cameras containing many rows of light-sensitive materials. When light from an object such as a machine part strikes the camera, the sensitive materials measure the intensity of light and convert the light rays into a range of numbers. The numbers are part of a grayscale system in which brightness is measured in a range of values. One scale ranges from 0 to 15, and another from 0 to 225. The 0 is represented by black. The highest number is white. The numbers in between represent different shades of gray. The computer then makes the calculations and converts the numbers into a picture that shows an image of the object in question. It is not yet known whether robots will one day have vision as good as human vision. Engineers and technicians believe they will, but only after years of development.

S: Are engineers and technicians working on other advanced designing and experimenting with new types of robots?

E: Certainly. Some engineers are writing new programs allowing robots to make decisions such as whether to discard defective parts in finished products. To do this, the robots will also have to be capable of identifying those defective parts. Recently, researchers at MIT introduced their latest robotic development affectionately known as Kismet (Turkish for good fortune) hailed as a toy of the future. The aim is to get further understanding of artificial intelligence and take us one step closer to the "thinking, seeing, learning" machine. The truly cognizant robot is

believed to be will be the norm in the future. Robotic surgeons can help during minimally invasive surgery. This improves patient recovery time, discomfort and possibly dangerous side effects. Nasa's Jet Propulsion Lab, Oak Ridge National Lab and the University of Southern California are collaborating to make mini-robot helpers for the battlefield—tiny, lightweight robotic warriors, known as microrovers. It is envisaged that the necessary movement will be provided by a unique wheel on struts. Each strut can be rotated 360° enabling the unit to stand tall, crouch low, right itself after flipping over, climb stairs and roll and somersault over rubber. Gathering and relaying information back will be by integrating visible and thermal infrared imagers and an acoustic or vibration sensor.

S: Are there any difficulties in developing new type of intelligent robots?

E: Sure. Despite a great deal of advances, one of the main hindrances to artificial intelligent life is the extreme complexity of human senses and our lack of real understanding in this area. For instance, how could we recreate a truly intelligent visual system, when we are not sure of all the parameters involved in human vision, how it works and interacts with other senses. Scientists may never truly discover the key to this conundrum or, if they do, find a way of transferring their findings to artificial life. These future robots, assembled with a sense of touch and the ability to see, to learn and make decisions, will have plenty of work to do. No one knows for sure what the descendants of robots such as Rodolph (equipped with three electrostatic transducers that can act as a transmitter of signals which bounce off objects rather like the squeals of bats or the clicking noises made by dolphins, or receivers for echoes) and Kismet are going to look like. One thing does seem certain, however, intelligent machines will one day be norm rather than the conception. Anyone wanting to understand the modern and future industry will have to know about robotics.

Notes

1. robots 机器人
2. prevalent 流行的，盛行的
3. are programmed and engineered to perform industrial tasks without human intervention 在无人介入的情况下，编入程序和设计执行工业任务
4. automotive industry 汽车工业
5. welding and spray painting automobile and truck bodies 进行焊接并给汽车和卡车车身喷漆
6. load and unload hot, heavy metal forms used in machines casting automobile and truck frames 装卸用于铸造汽车和卡车构架机器中高温和沉重的金属模板
7. electric motors 电动机
8. small appliances 小型家用电器
9. the radioactive materials 放射性物质
10. preventing human personnel from being exposed to radiation 防止人员受到辐射伤害

11. the reduction in job-related injuries 减少与工作有关的伤害
12. What makes a robot a robot? 是什么使机器人成为机器人？
13. doing spot welding 进行点焊
14. a sense of touch 触觉
15. engineers skilled in microelectronics 在微电子方面技能高的工程师
16. artificial vision for robots 机器人的人造视觉
17. inspect one specific class of objects out of a stack of different kinds of materials 从一堆不同种类的物质中检验出某一特定级别的物质
18. electronic digital cameras 电子数字照相机
19. many rows of light-sensitive materials 许多排光敏感物质
20. strikes 照射
21. the intensity of light 光的密度
22. convert the light rays into a range of numbers 将光线转变成一系列数字
23. The numbers are part of a grayscale system in which brightness is measured in a range of values. 这些数字是灰度装置的部分，根据一系列的数字可测出亮度。
24. one scale ranges from 0 to 15 从0到15为一个刻度范围（一个灰度值）
25. The numbers in between represent different shades of gray. 介于两者（黑白）之间的数字代表不同的灰度。
26. an image of the object in question 有关物体的图像
27. to discard defective parts in finished products 丢弃最终产品中有缺陷的部件
28. identifying those defective parts 鉴别出那些有缺陷的部件
29. MIT (the Massachusetts Institute of Technology) 麻省理工学院
30. affectionately known as Kismet 被亲切地称为 Kismet
31. Turkish for good fortune 土耳其语：好运
32. hailed as a toy of the future 把它称为将来的玩具
33. artificial intelligence 人工智能
34. the truly cognizant robot is believed to be the norm in the future 许多人相信真正有识别能力的机器人在将来会成为现实
35. Robotic surgeons can help during minimally invasive surgery. 机器人外科医生在小型的体内手术过程中能有帮助。
36. This improves patient recovery time, discomfort and possibly dangerous side effects. 这能缩短病人恢复的时间，减少痛苦和可能具有危险的负面影响。
37. Nasa's Jet Propulsion Laboratory 美国宇航局的喷气发动机实验室
38. Oak Ridge National Lab 国家实验室
39. collaborating to make mini-robot helpers for the battlefield 合作制造用于战场的微型机器人助手
40. tiny, lightweight robotic warriors 体积小、重量轻的机器人战士
41. microrovers 微型传感器

42. It is envisaged that ... 可以设想……
43. a unique wheel on struts 在支柱（撑杆）上的独特的轮子
44. enabling the unit to stand tall, crouch low 能使机器人站立，蹲下（匍匐）
45. right itself after flipping over 在跳跃后调整好自身的平衡
46. climb stairs and roll and somersault over rubber 爬梯子，在橡胶上打滚、翻筋斗
47. gathering and relaying information back 收集信息并向后传递
48. by integrating visible and thermal infrared imagers and an acoustic or vibration sensor 将可见的和热红外线图像显示器与一种有听觉或震动传感器结合
49. the main hindrances 主要障碍
50. complexity of human senses 人类知觉的复杂性
51. a truly intelligent visual system 一种真正的智能可视系统
52. when we are not sure of all the parameters involved in human vision 当我们还不能确信所有与人类视觉有关的参数时
53. It works and interacts with other senses. 它与其他知觉相互作用和相互影响。
54. this conundrum 这一复杂的难题
55. the descendants of robots 下一代机器人
56. equipped with three electrostatic transducers that can act as a transmitter 装配了具有发射器作用的三个静电传感器
57. signals which bounce off objects rather like the squeals of bats or the clicking noises made by dolphins 信号很像蝙蝠的尖叫声或海豚发出的咔哒噪声一样从物体反射
58. receivers for echoes 回声接收器
59. Intelligent machines will one day be norm rather than the conception. 在将来的某一天，智能机器会成为现实而不是一种构想。
60. robotics 机器人技术（机器人学）

15. The FMS—An Advanced Manufacturing Technique
（柔性制造技术——一种先进的制造技术）

　　Nick. V. Scheel, President & COO (Chief Operating Officer) of Ford Automobile Corporation, arrived at Changan Ford Automobile Corporation Ltd. on the afternoon of March 26, 2002. Mr. Green, his senior Engineer introduced the FMS to the managers and engineers of the Sino-US joint venture. This Hi-tech will be used to manufacture new cars, such as Fiesta, Mendeo and so on.

　　A: Comrades, today, Mr. Green, senior Engineer of Ford Automobile Corporation, will introduce us a new system of factory automation. It is said to be one of the major machine tool builders in the U.S., and has developed the flexible manufacturing system, which has been used to manufacture Buick car series by GM in Shanghai. Mr. Green is also an expert in this field. Several years ago the representative of Cincinati Milacron forwarded for his company a plan to the Shanghai Sub-council of the CCPIT about holding an exhibition of the FMS. The exhibition was approved and successful. I believe these kinds of activities will definitely enhance the mutual understanding and cooperation between the Chinese and American machine tool builders and car makers. Now, let's give a round of applause to express our hearty welcome to Mr. Green.

G: Ladies and gentlemen, during the past few years, machine tool makers—some in Europe, many in the U. S., even more in Japan—have begun to supply so-called flexible manufacturing system that herald something very close to the worker-less factory. FMS has completed a process of a factory automation that began in the 1950s. First came numerically controlled machine tools that performed their operations automatically according to coded instruction on paper or Mylar tape. Then came computer-aided design and computer-aided manufacturing, or CAD/CAM, which replaced the numerical control tape with the computer. The new system integrates all these elements. They consist of computer-controlled machine centers that sculpt complicated metal parts at high speed and with great reliability, robots that handle the parts, and remotely guided cars that deliver materials. The components are linked by electronic controls that dictate what will happen at each stage of the manufacturing sequence, even automatically replacing worn-out or broken drill bits and other implements.

A: Could you give us some ideas about the cost of a flexible manufacturing system?

G: Measured against some of the machinery they replace, flexible manufacturing system seems expensive. A full-scale system, encompassing computer controls, five or more machining centers, and the accompanying transfer robots, can cost $25 million. Even a rudimentary system built around a single machine tool, say, a computer-controlled turning center, might cost about $325,000, while a conventional numerically controlled turning tool would cost only $175,000. But the direct comparison is a poor guide to the economies flexible automation offers, even taking into account the phenomenal productivity gains and asset utilization rates that come with virtually unmanned round-the-clock operation. Because an FMS can be instantly reprogrammed to make new parts or products, a single system can replace several different conventional machining lines, yielding huge savings in capital investment and plant size. Flexible automation's greatest potential for radical change lies in its capacity to manufacture goods cheaply in small volumes. Today, seventy-five percent of all machined parts are produced in batches of 50 or fewer. Many assembled products, too, ranging from airplanes and tractors to office desks and large computers, are made in batches. In the past, batch manufacturing required machines to do single task. These machines had to be either rebuilt or replaced at the time of product change. Flexible manufacturing brings a degree of diversity to manufacturing never before available. Different products can be made on the same line at will. General electric, for instance, uses flexible automation to make 2,000 different

Part One
High-tech 高新技术

versions of its basic electric meter as its Somersworth plant in New Hampshire, with total output of more than one million meters a year.

A: Could I ask when and where flexible manufacturing systems were first developed?

G: Flexible manufacturing systems were developed in the U.S. more than a decade ago by Cincinnati Milacron Kearney & Trecker, and White Consolidated. The U.S. remains a world leader in the technology. The major machine tool builders are being joined by new supplies with great financial resources and technical abilities, such as GE, Westinghouse, and Bendix. However, most of the action in flexible automation is now in Japan.

A: Could you cite a few examples to illustrate the experience of the factories using FMS?

G: I'd like first to mention a plant, of the Fuji Complex near Mount Fuji in Japan. This plant, one of the two in the Fuji Complex, makes parts for robots and machine tools. The machining operation, occupying 54,000 square feet, is supervised at night by a single controller, who watches the machines on closed-circuit TV. The total cost of the plant is about $32 million, including the cost of 30 machining cells, which consist of computer-controlled machine tools loaded and unloaded by robots, along with material-handling robots, monitors, and a programmable controller to orchestrate the operation. The company estimates that it probably would have needed ten times the capital investment for the same output with conventional manufacturing. It also would have needed ten times its labor force of about 100. In this plant one employee supervises ten machining cells; the others act as maintenance men and perform assembly. All in all, the plant is about five times as productive as its conventional counterpart would be.

A: It is said that the most astonishing Japanese automated factory is near run by Yamazaki about 20 miles from its headquarters Nagoya and is described as a 21st century factory.

G: In that case, the plant's 65 computer-controlled machine tools and 34 robots are linked via a fiber-optic cable with the computerized design center back in headquarters. From there the flexible factory can be directed to manufacture the required types of parts as well as to make the tools and fixtures to produce the parts by entering into the computer's memory names of various machine tool models scheduled to be produced and pressing a few buttons to get production going. The Yamazaki plant is the world's first automated factory to be run by telephone from corporate headquarters. The plant has 213 men helping produce what would take 2,500 in a conventional factory. At maximum capacity the plant is able to turn out about $230 million of machine tools a year. But production is so organized that sales can be reduced to $80 million a year, if need be, without laying off workers. The Yamazaki plant illustrates another aspect—economy of scope: with flexible automation, a manufacturer can economically shrink production capacity to match lower market demand.

A: Mr. Green, I'd like to raise a more specific question. What else can robots do besides unloading raw metal blanks from the carriers cart and placing them in a lathe?

G: Besides the "pick and place" robots, there are assembly robots which put the parts together.

Moreover, the system may also include welding robots which take the job of joining the parts, making as many welds as necessary on all sides.

A: Could you tell us how inspection is done in a flexible manufacturing system?

G: The newly made product is scrutinized by a camera containing a semiconductor chip that can "see" and instantly measure deviations from standards.

A: Comrades. As it is already getting too late for Mr. Green, I assume that we ought to bring our discussion to a close. May I speak for all the Chinese comrades present here to thank Mr. Green for his sparing us so much time to introduce the FMS. We look forward to great success of our joint venture. Now, let us once more clap our hands to express our thanks to Mr. Green and his colleagues.

Notes

1. FMS (flexible manufacturing system) 柔性制造技术
2. forwarded for his company a plan to the Shanghai Sub-council of the CCPIT about holding an exhibition of the FMS 为他们公司向中国国际贸易促进会上海分会递交了一份有关举办一次柔性制造系统展览的计划
3. give a round of applause to express our hearty welcome to Mr. Green 让我们鼓掌，向格林先生表示热烈的欢迎
4. herald something very close to the worker-less factory 预示着几乎是无人工厂的来临。
5. FMS has completed a process of a factory automation that began in the 1950s. 柔性制造技术完成了始于20世纪50年代的工厂自动化的进程。
6. numerically controlled machine tools 数控机床
7. coded instruction on paper or Mylar tape. 在纸带或迈拉磁带上的编码指令
8. computer-aided design and computer-aided manufacturing, or CAD/CAM 计算机辅助设计/计算机辅助制造
9. They consist of computer-controlled machine centers that sculpt complicated metal parts at high speed and with great reliability, robots that handle the parts, and remotely guided cars that deliver materials. 它们包括以高速度和高可靠性雕刻复杂金属部件的计算机控制的机械中心，处理部件的机器人，传送物料的遥控车。
10. dictate what will happen at each stage of the manufacturing sequence, even automatically replacing worn-out or broken drill bits and other implements 指令在生产程序的每个阶段生产什么产品，甚至自动换掉破旧的钻头及其他工具。
12. measured against some of the machinery they replace 与它们所代替的一些机器相比较
12. accompanying transfer robots 附属的传输机器人
13. even a rudimentary system built around a single machine tool, say, a computer-controlled turning center 甚至围绕一台机床建立起的基本系统，例如，一个计算机控制的车削加工中心

14. But the direct comparison is a poor guide to the economies flexible automation offers, even taking into account the phenomenal productivity gains and asset utilization rates that come with virtually unmanned round-the-clock operation. 然而这种将柔性制造技术提供的经济效益进行直接的比较是一种误导，即使是考虑到24小时运转无人操作带来的超凡的劳动生产效益和很高的资产利用率。

15. Flexible automation's greatest potential for radical change lies in its capacity to manufacture goods cheaply in small volumes. 柔性制造技术根本变化的巨大潜力在于其小批量低成本生产商品的能力。

16. produced in batches of 50 or fewer 以50或更少的数量批量生产

17. Flexible manufacturing brings a degree of diversity to manufacturing never before available. 柔性制造技术带来了以前从未采用过的制造技术的多样化。

18. the major machine tool builders are being joined by new supplies with great financial resources and technical abilities 一些具有强大资金实力和技术能力的供应商也加入了大的机床制造商行业

19. Most of the action in flexible automation is now in Japan. 现在，柔性自动化方面的大部分加入行动发生在日本。

20. closed-circuit TV 闭路电视

21. consist of computer-controlled machine tools loaded and unloaded by robots, along with material-handling robots, monitors, and a programmable controller to orchestrate the operation 包括计算机控制的有装卸机器人的机床，物料处理机器人，检测装置和协调运转的程序控制器

22. probably would have needed ten times the capital investment for the same output with conventional manufacturing 用普通的制造技术生产相同产量的产品，可能需要10倍的投资

23. It also would have needed ten times its labor force of about 100. 它也将需要约1000个劳动力（约100个劳动力的10倍）。

24. The most astonishing Japanese automated factory is near run by Yamazaki about 20 miles from its headquarters Nagoya. 最令人惊讶的日本柔性制造技术工厂，距名古屋总部约20英里，即将由山奇公司经营。

25. via a fiber-optic cable 通过光线电缆

26. From there the flexible factory can be directed to manufacture the required types of parts as well as to make the tools and fixtures to produce the parts by entering into the computer's memory names of various machine tool models scheduled to be produced and pressing a few buttons to get producing going. 在那里，可以指令柔性制造技术工厂，将计划生产的各种型号的机床的名字输入计算机存储器，生产所需要的类型的部件以及生产机床和工具以制造部件，按下几个按钮就进行生产。

27. turn out about $230 million of machine tools a year 一年生产价值为2.3亿美元的机床

28. But production is so organized that sales can be reduced to $80 million a year, if need be, without laying off workers. 但生产如此有组织，以至于把销售量降到8千万美元，如果需

要，可不临时解雇工人。

29. economy of scope 规模经济
30. A manufacturer can economically shrink production capacity to match lower market demand. 生产厂商可以经济地缩小生产量以适应较低的市场要求。
31. besides unloading raw metal blanks from the carriers cart and placing them in a lathe 除了从载车上卸下金属原材料并将它们放到车床上外
32. "pick and place" robots 装卸机器人
33. making as many welds as necessary on all sides 在所有工作面尽可能地进行焊接
34. The newly made product is scrutinized by a camera containing a semiconductor chip that can "see" and instantly measure deviations from standards. 新产品由一台装有半导体芯片的照相机仔细检查，它能看见并立即测出与标准的误差。
35. bring our seminar to a close 结束我们的研讨会

16. Discussion of Manufacturing Technological Problems（1）
（讨论制造技术问题之一）

A：Could you explain the basic concept for projecting the new hot mill line?

B：OK. This project of modernization will bring notable benefits. First, production rate will be raised by three times, which will enrich a solid foundation for development of your plant. The quality of hot rolled products will be improved sharply, and some indices of quality will come up to an international advanced level. The finished product rate will rise while production cost will be reduced. You saw the hot and cold mills yesterday. Could you tell us your tentative ideas about modernization?

A：All right. After we saw the mills, we found that most of parts of both hot and cold mills can be reused. Only a small part of them should be replaced or changed. What parts must be replaced or changed? Would you explain the items to be modernized for the hot mill first?

B：With pleasure. The existing work roll and it's chocks, coolant headers and side guides must be replaced.

A：Please tell us something about emulsion spray and control equipment.

B：Emulsion spray headers will be mounted at the two sides of top and bottom rolls. Each spray header has three rows of nozzles. They spray roll brushes and the roll gap and cool the work rolls. Nozzles can be independently controlled by hand or by an auto switch. They can also be controlled in groups.

A：Will the top back-up rolls be equipped with spray headers?

B：OK. Each nozzle will be remotely controlled by a solenoid valve.

A：What will be used to control the flow rate of emulsion?

B: It will be controlled by the pressure adjusting valve from the main inlet pipe.

A: I see. Please go ahead.

B: The top roll balance should be changed. In order to widen the opening of the work roll, the existing chocks of back-up rolls must be machined.

A: Machining work will be done by us. I hope you will supply drawings for machining.

B: No problem. The work roll brushing device must be added to the hot roughing mill.

A: What is included in the work roll brushing device?

B: Includes the brush rolls made of steel wire, stands of brush rolls, hydraulic motors for rotating and driving and hydraulic motors for oscillating.

A: I understand.

B: Let's discuss the items for changing the cold mill into a hot finishing mill.

A: Fine.

B: Most of the items to be modernized for the cold mill are the same as those of the hot mill.

A: What items are different?

B: I'm coming to that. The existing electro-mechanic screw-down must be changed into a hydraulic mill. The top covers of the great reducer of screw-down and flanges connected to the screw should be mating machined with original equipment. The machining work will be done by you.

A: OK. We hope you provide drawings and the detectors of the screw-down position.

B: But I think my company will supply drawings and the detectors will be supplied by the Electrical Company.

A: That's fine.

B: The existing entry and exit tension coiler and roll coolant system should be replaced. The spacing and number of nozzles of the coolant system are different from those of the hot mill.

A: I see. We want to add a belt wrapper, the systems of shape control and automatic gauge control on the new hot mill line.

B: Very good. I think we should also add a side trimmer, scrap conveyer, 3-roll bender, banding tool and radiation isotope gauges.

A: How many radiation isotope gauges will you supply?

B: We will supply two. They will be arranged at both the entry and exit sides of the finishing mill.

Notes

1. the basic concept for projecting the new hot mill line 规划新的热扎生产线的基本思想
2. enrich a solid foundation 奠定坚实的基础
3. indices of quality 质量指标
4. the finished product rate 产品的成品率
5. the hot and cold mills 冷轧机和热轧机

6. tentative ideas 初步设想
7. the existing work roll and it's chocks, coolant headers and side guides 现存的工作辊及其轴承座，冷却液集管和测导尺
8. emulsion spray 乳液喷射
9. Emulsion spray headers will be mounted at the two sides of top and bottom rolls. 乳液集管安装在上下轧辊两侧。
10. Each spray header has three rows of nozzles. 每个喷射集管有三排喷嘴。
11. They spray roll brushes and the roll gap and cool the work rolls. 喷嘴向清刷辊和扎辊间隙喷乳液并冷却工作辊。
12. auto switch 自动开关
13. Will the top back-up rolls be equipped with spray headers? 上支承辊要装配喷射集管吗？
14. a solenoid valve 电磁阀
15. the pressure adjusting valve from the main inlet pipe 总进液管的调压阀
16. the top roll balance 上轧辊平衡
17. to widen the opening of the work roll 加大工作辊的开度
18. the existing chocks of back-up rolls must be machined 必须对现有的支撑辊轴承座进行加工
19. machining work 加工工作
20. The work roll brushing device must be added to the hot roughing mill. 热粗轧机上必须增设工作辊装置。
21. the brush rolls made of steel wire, stands of brush rolls, hydraulic motors for rotating and driving and hydraulic motors for oscillating 钢丝刷清刷辊、清刷辊辊座、旋转驱动用液压马达和震颤用液压马达
22. the items for changing the cold mill into a hot finishing mill 冷轧机改造成热精轧机的项目
23. I am coming to that. 我马上就讲到。
24. The existing electro-mechanic screw-down must be changed into a hydraulic mill. 应将现有的电动机械压下机构改为液压下机构。
25. The top covers of the great reducer of screw-down and flanges connected to the screw should be mating machined with original equipment. 压下机构的齿轮减速机的顶盖和与压下螺丝连接的法兰应与原设备配套加工。
26. the detectors of the screw-down position 压下位置检测器
27. the existing entry and exit tension coiler and roll coolant system 现有的进、出口张力卷取机和轧辊冷却液系统
28. the spacing and number of nozzles of the coolant system 冷却液系统的喷嘴间距与数量
29. to add a belt wrapper, the systems of shape control and automatic gauge control on the new hot mill line 在新的热轧生产线上增设带式助卷器，扳形控制和厚度自动控制系统。
30. side trimmer, scrap conveyer, 3-roll bender, banding tool and radiation isotope gauges 切边剪、废料运输机、三辊弯辊装置、打捆工具和辐射式同位素测厚仪

17. Discussion of Manufacturing Technological Problems（2）
（讨论制造技术问题之二）

B: Could you explain the temperature measuring device?

A: My pleasure. We'll provide five temperature measuring sensors.

B: What's the type of those sensors?

A: Two contact types, two radiant types, and one manually operated type.

B: I think the new hot mill line need four guillotine shears. We want to keep the existing two shears and add one heavy crop shear and one hydraulic down cut shear.

A: That's good.

B: Could you explain the drive of the heavy crop shear?

A: The drive are composed of a fly wheel, pneumatic friction clutch, brake, motor coupling and a triple reduction gear unit. Gears are helical type and made of alloy steel, hardened and ground. The internal part of the reduction gear will be oil splash lubricated.

B: I understand. Thank you. Let's discuss the emulsion circulating system.

A: Fine.

B: If you confirm that the existing emulsion circulating system can meet requirements for the new flow rate and pressure, we'll not change it.

A: It's hard to say yes or no. Before I make a conclusion, I need some calculations and your drawings and information.

B: We'll provide you with drawings and information related to the emulsion system. Will you make a conclusion before you leave here?

A: No, it's time consumable to make this calculation. I can't finish it before we leave. I suggest we decide to change or keep the emulsion system next time.

B: OK. How do you consider the automation after modernization?

A: For automation of a single machine or a unit, one or more programmable logic controllers will be used to carry out analog and digital process. Control logic is accomplished with switch selection. These controllers can be set manually or automatically. We'll provide a complete set of hardware, such as programmable controller, input, output, display, printer and cables. We'll provide application software for this set of device, too. In order to obtain some important data in production process, we'll also furnish some necessary inspection and measuring devices, such as thickness gauge, pressductor, temperature measuring device, position transducer, limit switch etc. so as to accomplish open and closed loop control of controlled targets. Without a doubt, we can provide a higher-level computer to control the production process of the whole rolling line if you need, but investment cost will increase.

B: Well, we'll think about it.

A: That's all for our tentative ideas about modernization. What we said are just superficial ideas. Since the project is very big and will involve much work and technical details, I must have a detailed discussion with other experts after I return to America and make a reasonable plan of modernization.

B: We hope you make a ripe and complete plan.

A: Sure, we will.

B: What's your plan for arrangement of a schedule for this project?

A: We'll make an FOB delivery in eighteen months from the date when the contract come into force, including packing, transportation and shipment of the contract equipment, since it is a very complex and formidable work to modernize the existing equipment, so within one month after signing the contract, we'll send three to five engineers to the plant site for investigating, surveying and drawing the information and data related to the modernization. We'll finish the preliminary design of the whole line within three months. We request that you inspect, confirm and sign the preliminary design. After that, we will carry out a detailed design and it will be inspected by you, too. We'll submit some of the drawings to your Chinese co-manufacturer.

B: What stage will we undergo for handing over the completed project?

A: The construction of foundation should be done first and then erection of equipment will be carried out. Our technical personnel will give you instruction of erection. After erection, a test runs without and with a load will be performed. After trial production for some time, the final acceptance test of products will be done and the main indices specified in the contract, such as indices of production rate and quality will be checked and specified in detail. The acceptance shall be the final sign and termination of the contract.

Notes

1. temperature measuring sensors 测温传感器
2. two contact types, two radiant types, and one manually operated type 两个接触式，两个辐射式，一个手动操作式
3. guillotine shears 剪断机
4. shears 剪切机
5. one heavy crop shear and one hydraulic down cut shear 一台重型切头剪和一台液压下切剪
6. fly wheel, pneumatic friction clutch, brake, motor coupling and a triple reduction gear unit 飞轮、气动摩擦离合器、抱闸、电机连轴节和三级减速剂组成
7. Gears are helical type and made of alloy steel, hardened and ground. 齿轮呈螺旋形，用合金钢制造，经淬硬和研磨处理。

8. The internal part of the reduction gear will be oil splash lubricated. 减速机内部将用稀油飞溅润滑。
9. the emulsion circulating system 乳液循环系统
10. programmable logic controllers 可编程序逻辑控制器
11. analog and digital process 模拟和数字处理
12. Control logic is accomplished with switch selection. 通过开关选择，实现逻辑程序控制。
13. programmable controller, input, output, display, printer and cables 编程控制器、输入、输出、显示、打印和电缆
14. thickness gauge, pressductor 测厚仪，测压头
15. position transducer, limit switch 位置传感器和限位开关
16. accomplish open and closed loop control of controlled targets 对控制对象实现开环或闭环控制
17. to control the production process of the whole rolling line 实现对整个轧制生产线生产过程的控制
18. superficial ideas 肤浅的看法
19. a ripe and complete plan 一个成熟和完善的计划
20. arrangement of a schedule for this project 工程的进度安排
21. very complex and formidable work 复杂而艰巨的工作
22. for investigating, surveying and drawing 调查与测绘
23. preliminary design of the whole line 总体的初步设计
24. submit some of the drawings to your Chinese co-manufacturer 将一部分图纸呈给中国合作制造商
25. the construction of foundation 基础施工
26. After erection, a test runs without and with a load will be performed. 安装完后，将进行无负荷和有负荷试车。
27. trial production 试生产
28. the final acceptance test of products 最终的产品验收实验
29. such as indices of production rate and quality will be checked and specified in detail 如产量和质量指标，将分别进行考核和详细规定
30. The acceptance shall be the final sign and termination of the contract. 这项验收是本合同的最终标志和终止。

Additional Expressions
单独（集体）转动 individual (group) drive
中间辊道 intermediate table
拨料头装置 crop end push-off device
料尾推出装置 pusher for crop ends
换刀刃装置 knife changing device
刀刃倾角　knife rake

高新技术、技术转让与国际工程合作

定位卡板　keeper plate
回转升降台　turnstile lifting table
阀座，阀架　valve stand
伸缩装置　retracting device
四列圆柱滚子轴承　4-row cylindrical roller bearing
双列圆锥滚子轴承　double-row tapered roller bearing
抱卷小车　coil gripper
存卷小车　coil storage car
夹送辊/导向辊　pinch/deflector roll
乳液吹净装置　emulsion blow-off device
伞齿轮　bevel gear
地脚板　bedplate
推杆　ram
储势器　accumulator
曲轴　crank shaft
轴向柱塞泵　axial piston pump
液压上推控制器　hydraulic push-up controller
摆动液压缸　traverse cylinder
星行碎边机　star type scrap chopper
力矩　moment
转矩限制器　torque limiter
机架，牌坊；壳，罩　housing
卷材抓取器　coil grab
轧制规程　rolling schedule
连续可变凸度　continuous variable crown (CVC)
辊缝内部控制装置　ingap gauge control system (IGC)
厚度自动控制　automatic gauge control (AGC)
油雾润滑　oil mist lubrication
偏心导筒　accentric sleeve
废料收集装置　scrap collector
辊式助卷器　roller wrapper
公用设施　utility
平台　platform
走道　walkway
地板　floor plate
地沟盖板　trench cover
电器布线材料　electrical wiring material
外置（外伸）轴承　outboard bearing

刀架　knife carriage
横梁　crossbeam
撇污装置　skimmer
格栅　grid
单轨　monorail
灭火系统　fire extinguishing system
外形尺寸　outline dimension
辊缝调整　roll gap adjustment
扎制道次　pass
压下量　reduction
每道次压下量　reduction per pass
压力传感器　load cell
侧导卫板　side guard
配重缸，抗衡缸　counter balance cylinder
双锥形辊　double tapered roller
链条和链轮　chain and sprocket
开卷刮板　peeler
来料镰刀弯　camber
卷轴真圆直径　true mandrel diameter
扇形块　segment
展平辊　crossbow roll
导筒斜台　sleeve ramp
叉车　fork lift
挠性联轴节，弹性联轴节　flexible coupling
包箍辊；外卷辊　wrapper roll
中间合金　master alloy

18. Environmentally Friendly Farming in the U. S. A.
（美国的环境友好型农业）

　　Hendler (H), who is studying agriculture and natural resources in South Dakota State University, and Prof. Korant (K) are discussing the problem of "Environmentally Friendly Farming".
H: Excuse me, Prof. K, I want to ask you a question on the agriculture of the U. S. A.
K: Go ahead!
H: Are U. S. farmers planting more acres of crops using soil building and pollution fighting farming

systems than traditional methods that rely on the plow or intensive tillage?

K: Recently I have read a report, titled *National Crop Residue Management Survey*, which shows a 6 million acre grain for environmentally friendly farming systems, and traditional farming methods which result in greater soil erosion and runoff from fields, declined by 4 million acres this year. The survey, conducted on a county-by-county basis by USDA Natural Resources Conservation Service, indicates that farmers in Iowa, Illinois, South Dakota, Kansas, and Indiana contributed the most to the increase in acres grown with environmentally friendly farming systems known as conservation tillage systems, accounting for 5 million of the 6 million acre increase in conservation tillage in 2000.

H: What are the tillage techniques of the conversation tillage systems in the United States?

K: All conservation tillage systems, such as no-till, mulch-till, ridge-till, strip-till, and zone-till, rely on less tillage or less soil disturbance to plant and manage crops. Farmers who use these systems leave plant materials—leaves, stems, roots, etc. on the surface of fields after harvests, which serve as a blanket to protect the soil from erosion. The crop residues slowly decompose to add organic substance to the soil, much like mulching or composting to add organic matter to a garden.

H: How are the conservation tillage systems applied to the farming in the U.S.A.?

K: The results of the investigation for 1997 indicate that conservation tillage systems now account for 109.8 million acres or fully 37% of the 294.6 million annually planted cropland acres in the United States, and traditional systems that rely on the plow or intensive tillage fell to 107.6 million acres in 2000. The remaining acres are in an intermediate farming system known as reduced-till. The head of the nonprofit center that compiles and publishes the annual survey is calling on consumers and farmers alike to focus increased attention on conservation tillage systems.

H: What are the other advantages that have been apparently demonstrated by using conservation tillage systems in the U.S.A. in recent years?

K: This is a good question. "Independent research and practical application across the country show that these systems not only replenish and build organic fertilizer in the soil for improved future food productivity but they will also protect water quality and enhance the number of wild animals and the environment for future generations. There is also growing evidence that these systems can even help us combat the potential for global warming. Conservation tillage has long been credited for protecting water quality by reducing runoff from farm fields," says John Hebblethwaite, executive director of the Conservation Technology Information Center. The latest research also indicates that soil enriched by crop residues offers natural protection for groundwater. Conservation tillage systems have saved the farmer money by reducing the use of farming machines and trips through the field for planting and cultivation.

Notes

1. Environmentally Friendly Farming in the U. S. A. 美国的环境友好型农业
2. intensive tillage 密集耕耘
3. *National Crop Residue Management Survey* 全美庄稼残留物管理调查
4. soil erosion and runoff from fields 土壤侵蚀和水土流失
5. USDA（U. S. Department of Agriculture）Natural Resources Conservation Service 美国农业部自然资源保护局
6. conservation tillage systems 免耕法（保留式耕作法）
7. no-till, mulch-till, ridge-till, strip-till, and zone-till 休耕法（无为耕作法）、覆盖耕作法、垅行耕作法、带状耕作法和区域耕作法
8. mulching or composting 用覆盖料或堆肥
9. an intermediate farming system known as reduced-till 叫做减耕的中间耕作法
10. independent research and practical application across the country 独立的研究和在全美的实际应用
11. replenish 补充
12. enhance the number of wild animals and the environment for future generations 增加野生动物的数量，改善环境，造福后代
13. help us combat the potential for global warming 有助于防止全球变暖的趋势
14. "Conservation tillage has long been credited for protecting water quality by reducing runoff from farm fields." 免耕法因能减少农田土壤流失而保护水质一直受到赞赏

19. Cloning
（克隆）

A: Hi, what are you reading?
B: I'm reading a Chinese classic novel.
A: What's the title?
B: I'm reading *Pilgrimage to the West*.
A: Oh, that's a very popular novel in China. The Monkey King is the hero.
B: Yes, but I'm thinking of another thing while reading, when the Monkey King could pull out a piece of fine hair from its body and with a magic wind from its mouth, the hair could be changed into numerous imitations of Monkey King. I think this is the earliest creative idea of cloning.
A: Wonderful! In this way clone sheep Dolly might be an "infringement" of an ancient Chinese creative idea.

B: Could we do anything about it?

A: I'm afraid not, because the Monkey King and the novel is only a legend and our ancestors did not patent this idea.

B: But there is such a story. I once read an article which says the German chemical giant, BASF, was once refused a patent for the clever idea of pumping expanded plastics into a submerged ship and thereby floating it to the surface. The reasons of the refusal were that the German Examiner had once seen a Walt Disney cartoon in which Donald Duck had performed a similar trick on a sunken boat with table-tennis balls. If the BASF scheme proves successful in practice and enables valuable wrecks to be salvaged, it is likely that Walt Disney will be credited as the inventor.

A: That's very interesting. Anyway, cloning technology is really a wonder. But I'm still not quite clear how to clone.

B: I know little about it. It's said in July 1996, a team of Scottish scientists produced the first live birth of a healthy sheep cloned from an adult mammal. And the embryo developed normally and was delivered safely. Named Dolly, this healthy sheep was introduced to the world with much fanfare in February 1997.

A: Is Dolly similar to that adult mammal?

B: Of course. Dolly has most of the genetic characteristics of the adult sheep. But the young Dolly has no father.

A: What kind of biotechnology was used to create Dolly by Wilmut and his colleagues?

B: They concentrated on arresting the cell cycle—the series of choreographed steps all cells go through in the process of dividing. In Dolly's case, the cells the scientists wanted to clone came from the udder of a pregnant sheep. To stop them from dividing, researchers starved the cells of nutrients for a week. In response, the cells fell into a slumbering state that resembled deep hibernation.

A: Did Wilmut and et al. switch to mainstream cloning technique known as nuclear transfer immediately after that?

B: Yes. They first removed the nucleus of an unfertilized egg, while leaving the surrounding cytoplasm intact. Then they placed the egg next to nucleus of a quiescent donor cell and applied gentle pulse of electricity. These pulses prompted the egg to accept the new nucleus—and all the DNA it contained—as though it were its own. They also triggered a burst of biochemical activity, jump-starting the process of cell division. A week later, the embryo that had already started growing into Dolly was implanted in the uterus of a surrogate ewe.

A: Could the exotic form of reproductive engineering become an extremely useful tool?

B: Definitely. The ability to clone adult mammals, in particular, opens up myriad exciting possibilities, from propagating endangered animal species to producing replacement organs for transplant patients. Agriculture stands to benefit as well. Dairy farmers, for example, could clone their champion cows, making it possible to produce more milk from smaller herds. Sheep ranchers could do the same with their top lamb and wool producers. Cloning could also provide an efficient way of creating flocks of sheep that have been genetically engineered to produce milk laced with valuable enzymes and drugs. A potential treatment for cystic fibrosis is among the pharmaceuticals PPL is looking at. The creation of Dolly represents a unique advance for cloning technology. Rather than a prelude to, however, many scientists herald the achievements as the forerunner of a revolution in animal breeding.

A: Are there any difficulties and problems concerning the clone technique?
B: Sure. It's easy to imagine the technology being misused, and as news from Roslin spread, apocalyptic scenarios proliferated, journalists wrote seriously about the possibilities of virgin births, resurrecting the dead and women giving birth to themselves. If cloning were perfected, there would be no need for men. Like most of the scientists who score major breakthroughs, Wilmut et al. have raised more questions than they have answered. Among the most pressing are questions about Dolly's health. She came from a six-year-old cell, she had exhibited signs of aging prematurely before she died in early 2003. In addition, as the high rate of spontaneous abortion suggests, cloning sometimes damages DNA. As a result, cloned animals may develop any numbers of diseases that will shortened their lives. Indeed, cloning an adult mammal is still a difficult, cumbersome business, so that even agriculture and biomedical application of the technology could be years away.
A: If someone researches into human cloning, that will be terrible.
B: That's a serous problem. It has inevitably intensified the debate about subjecting humans to cloning. And this is probably a good thing because it gives the public time to digest clone technique and policymakers time to find ways to prevent abuses without blocking scientific progress.
A: That's a good idea.
B: I hope we may clone more and more pandas in the future.
A: I agree with you. Oh, I forget to tell you I get two tickets for a football match. Let's go.
B: Let's go.

Notes

1. *Pilgrimage to the West*《西游记》
2. imitations of Monkey King 克隆孙悟空
3. In this way clone sheep Dolly might be an "infringement" of an ancient Chinese creative idea. 这么说，克隆羊多利就是对中国古代一个创意的"侵权"了。
4. legend 传说
5. ancestors 祖先
6. the German chemical giant, BASF 德国大型化工厂：巴赛夫
7. pumping expanded plastics into a submerged ship and thereby floating it to the surface 向沉船

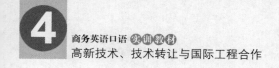

泵入膨胀塑料使其浮上水面

8. Walt Disney cartoon 沃尔特·迪斯尼卡通片
9. Donald Duck had performed a similar trick on a sunken boat with table-tennis balls. 唐老鸭表演了一个类似的花招——用一些乒乓球使沉船浮上水面。
10. enables valuable wrecks to be salvaged 能够将贵重的沉船打捞起来
11. It is likely that Walt Disney will be credited as the inventor. 有可能沃尔特·迪斯尼将作为发明人而要求赔。
12. A team of Scottish scientists produced the first live birth of a healthy sheep cloned from an adult mammal. 一个苏格兰科学家小组从成年哺乳动物克隆的世界上第一只健康的活羊诞生了。
13. the embryo developed normally and was delivered safely. 胚胎正常生长发育并安全分娩。
14. much fanfare 巨大轰动
15. the genetic characteristics 基因（遗传）特征
16. concentrate on arresting the cell cycle 集中掌握了抑制细胞的发育周期
17. the series of choreographed steps all cells go through in the process of dividing 在细胞的分裂过程中，所有细胞都经历的一系列仔细设计的步骤
18. the udder of a pregnant sheep 一只怀孕母羊的乳腺
19. starved the cells of nutrients 停止给细胞供给营养
20. In response, the cells fell into a slumbering state that resembled deep hibernation. 引起的反应是：细胞则处于类似深度冬眠那样的休眠状态。
21. Wilmut and et al. 威尔马特及其同事
22. switch to mainstream cloning technique known as nuclear transfer 转而采用被称为核转移的主流克隆技术
23. the nucleus of an unfertilized egg 一个未受精卵中的细胞核
24. leaving the surrounding cytoplasm intact 使周围的细胞质完好无损
25. Then they placed the egg next to nucleus of a quiescent donor cell and applied gentle pulse of electricity. 然后他们把该卵细胞放置在休眠的供体细胞旁边，并施以轻微的电脉冲。
26. These pulses prompted the egg to accept the new nucleus. 这些电脉冲激发卵细胞接受新的细胞核。
27. as though it were its own 就好像它是自己的一样
28. They also triggered a burst of biochemical activity, jump-starting the process of cell division. 电脉冲同时激发了生化化学活度，触发细胞分裂过程。
29. The embryo that had already started growing into Dolly was implanted in the uterus of a surrogate ewe. 开始生长为多利的胚胎被移植进代乳母羊的子宫中。
30. exotic form of reproductive engineering 这种异乎寻常的繁殖工程形式
31. opens up myriad exciting possibilities, from propagating endangered animal species to producing replacement organs for transplant patients 开创了无数的令人兴奋的可能性，从繁殖濒危动物的物种到为需要移植器官的病人生产可以替代的器官

32. stands to benefit 一定受益
33. dairy farmers 牛奶厂的工人们
34. champion cows 一流的奶牛
35. smaller herds 较小的牛群
36. sheep ranchers 牧羊场主们
37. with their top lamb and wool producers 最好的产羔羊和产毛羊
38. have been genetically engineered to produce milk laced with valuable enzymes and drugs 使这些羊经过遗传工程处理而生产出含有用的酶和药物成分
39. A potential treatment for cystic fibrosis is among the pharmaceuticals PPL is looking at. 而对囊性纤维变性具有潜在治疗作用的药物正是PPL公司所关注的众多药物之一。
40. The creation of Dolly represents a unique advance for cloning technology. 多利的诞生标志着克隆技术的一次空前飞跃。
41. rather than a prelude to 这不仅是一个序幕
42. many scientists herald the achievement as the forerunner of a revolution in animal breeding 许多科学家预言：该成就是动物饲养中一场革命的先兆
43. being misused 被滥用
44. as news from Roslin spread 当罗斯林研究所的消息传开时
45. apocalyptic scenarios proliferated 出现了大量描写世界末日恐怖景象的电影剧本
46. virgin births, resurrecting the dead and women giving birth to themselves 处女生育，使死者复活，妇女自己生产自己
47. score major breakthroughs 取得重大突破
48. signs of aging prematurely 过早地显示老化迹象
49. as the high rate of spontaneous abortion suggests 如同高比例自然流产所显示的一样
50. develop 患上
51. cumbersome business 麻烦的事
52. could be years away 数年以后
53. It has inevitably intensified the debate about subjecting humans to cloning. 这不可避免地已使有关将人类克隆的问题的争论变得更为激烈。
54. digest clone technique 领悟克隆技术
55. find ways to prevent abuses without blocking scientific progress 在预防滥用方面找到办法而又不阻碍科学的进步

20. The Development and Prospect of Genetically Modified Organism Crops in the World
（世界转基因农作物的发展和前景）

Wang Dong (W), a university graduate who is studying biological engineering and his supervisor, Prof. Gao Yuan (G) are discussing transgenic crops in the globe.

W: It is said that the scientists of our country have cultivated and planted a new kind of cotton, named genetically modified cotton (also called Monsanto's BT cotton or BT cotton), Prof. G, would you please give us a brief introduction to BT cotton?

G: Well, BT cotton is engineered to contain the bacterium Bacillus Thuringiensis (BT), which produces a toxin that kills bollworms, a major cause of crop losses. BT cotton has increased U.S. crop yields by 8% - 10% and boosted crops in China and other less-developed nations by as much as 50%.

W: Why has China researched and planted BT cotton?

G: Everyone knows that China, with over 1.3 billion people to feed, faces crop yields that lag far behind those of the developed world. China planted more than one million hectares of genetically modified cotton during the past year to improve crop yields, and is likely to boost that amount in the next planting season. China's adoption of Monsanto's BT cotton for about one-fourth of its crop is the evidence that the use of genetically modified plants will continue to grow despite European opposition. "Biotechnology is high on China's priority list," "We expect to see them continue to expand in this area," and "What really impresses me is not only the overall amount of acreage, but the fact that some 600,000 Chinese farmers decided to plant BT cotton," Robert Fraley, co-president of Monsanto's agricultural sector, told Reuters in an interview.

W: Why has China declined to comment on the nation's policy toward transgenic crops?

G: One agriculture ministry official recently said that it was a "state secret".

W: What about the BT cotton in the U.S.A.?

G: Fraley, the pioneer of genetically modified organisms—known in the industry as GMOs, and a team of three other Monsanto scientists were in Washington to receive the national medal of technology from President Clinton. They would be recognized for two decades of research that resulted in Roundup Ready Soybeans, which would be planted on half of the 75 million U.S. soybean acres in 2000 according to analysts. BT cotton was likely to be used on more than half the U.S. cotton acreage planted in 1999. U.S. farmers' swift adoption of GMOs is the evidence that the products are here to stay, despite complaints from environmental groups.

W: Are there any problems or troubles about the new GMOs crops in the world?

G: Yes, there are. The European Union has been slow to approve new GMOs crops, under pressure from activists who contend that health risks have not been fully researched; several U. S. grain processors said earlier this month that they would not buy certain types of GMOs corn from U. S. farmers until the types were approved by the European Union.

W: As I know, some experts have their comments on new GMOs crops, such as, "These are the most carefully studied plants ever to exist in agriculture," said Stephen Rogers, head of biotechnology projects in Monsanto's European division. "Each of more than 30 countries have conducted studies on the food and feed safety, environmental and pesticide risks of these plants, and they have been found completely safe." Prof. G, I am still worried about the sales of new GMOs crops in the markets presently.

G: Fraley suggested that labeling of GMOs products and food is probably to be "one of the solutions" to open up the European market. "The free flow of trade in agriculture is important. We cannot end up with one trading bloc controlling these products", he said.

W: Do the new GMO crops have an encouraging prospect?

G: I think so. Plant researchers envision designer crops in the future that will easily withstand drought, produce healthier cooking oils, provide more nutrients to animals, and manufacture enzymes and proteins for use in human drugs. "The level of innovation is breathtaking Fraley said. "Monsanto alone expects to roll out at least 40 new products over two or three years." Among the new products will likely be wheat engineered to resist fungal diseases. However, GMOs wheat has lagged behind corn, tomatoes, and other crops because it carries three copies of each gene and is more difficult to engineer. "We're very optimistic about the opportunities with wheat", said Fraley. Plant researchers at Monsanto and other companies hope soon to leap beyond genetic engineering that simply builds an better resistance to pests or diseases.

W: Today I have learned much knowledge of GMOs. Prof. G, thanks a lot.

G: You are welcome.

Notes

1. Genetically Modified Organism Crops (GMOs) 转基因农作物
2. supervisor 导师
3. transgenic crops 转基因作物
4. Monsanto 蒙桑托（葡空军基地）
5. bacterium Bacillus Thuringiensis (BT) (BT) 杆菌
6. a toxin 毒素
7. bollworms 棉铃虫，蟒蚣
8. boosted 增加
9. on China's priority list 中国优先发展的项目

10. overall amount of acreage 总面积的数量
11. Fraley, co-president of Monsanto's agricultural sector 蒙桑托农业部门两总裁之一，弗雷利
12. Reuters 路透社
13. state secret 国家机密
14. to receive the national medal of technology 接受国家技术奖章
15. Roundup Ready Soybeans 易于密植的大豆
16. swift adoption of GMOs 迅速种植转基因农作物
17. are here to stay 被普遍接受
18. Stephen Rogers, head of biotechnology projects in Monsanto's European division 在蒙桑托欧洲分部生物工程的负责人，斯蒂芬·罗杰斯
19. labeling of GMOs products and food 给转基因产品和食品贴上标签
20. We cannot end up with one trading bloc controlling these products. 我们不能只与一家控制这些产品的商贸集团结束交易。
21. envision designer crops 预想由专门设计师设计的农产品
22. withstand drought 耐旱
23. nutrients 营养物
24. enzymes and proteins 酶和蛋白质
25. The level of innovation is breathtaking. 创新水平令人吃惊。
26. to roll out 推广
27. engineered to resist fungal diseases 设计防止病菌
28. optimistic 乐观的
29. leap beyond genetic engineering 超越基因工程
30. an better resistance to pests or diseases 更有效防止病虫害

21. Zhuhai International Aviation and Aerospace Exhibition (ZIAAE)
（珠海国际航空航天展）

S—a student from Engineering Department   E—a senior engineer

S: Does ZIAAE indicate that China has become one of the world's few countries with a complete aviation industry?

E: I think so. Since William E. Boeing and his associate, Conrad Westervelt, finished the first B & W plane in America in 1916, the first airplane ever made in China took off on its trial flight at the airport of the Dashatou Aviation Bureau in Guangdong on Aug. 9, 1923. The event attracted even Dr. Sun Yat-sen, then the leader of the Guandong-based Revolutionary Army, and his wife Madame Soong Ching Ling, who also had their presence at the trial flight

Part One
High-tech 高新技术

photographed. On Nov. 5, 1996, the first international aviation and aerospace exhibition ever held in China opened to great fanfare, again in Guangdong Province, this time in the city of Zhuhai, a special economic zone over kilometers from Guangzhou and near the former residence of Sun Yat-sen. This most recent event demonstrates that China has entered the world's few countries with a complete aviation industry.

S: Could you give us a brief introduction to the history of China's aviation industry?

E: No problem. China's aviation industry actually started after the founding of the new China in 1949. In the beginning China could only repair airplanes and make aircrafts on license. But the country has gradually developed the capability of designing and making airplanes, and even making and launching man-made satellites. Today, China is able to design and manufacture a variety of planes, including fighters, bombers, attack planes, helicopters, air ferries and civilian passenger planes. Since 1951, China has altogether manufactured over 14,000 military and civil planes and over 50,000 aviation engines of various designs.

S: What about the development of China's civil aviation industry?

E: Very rapidly. Since the 1980s, the number of civilian planes in China has multiplied. By the end of Oct. 1996, the total number of passenger planes in the country amounted to 437, more than double the number in 1990, which was 207. Presently, China's air transport continues to increase by average of 20 percent annually; the rank of total transport volume in China's civil aviation industry compared to other members of the International Civil Aviation Organization has risen from 33rd in 1980 to 11th; and the rank for passenger volume has risen from 33 rd to 7th. Meanwhile, China's aviation and aerospace industry have been developing at lightning speed.

S: I've heard that for many years. China's aviation and aerospace industries have been planning to host an international airshow to enhance exchanges with foreign countries and promote the trade of China's aviation and aerospace technologies. Did ZIAAE finally turn the country's long-cherished dream into reality?

E: Certainly. The first Chinese airshow attracted over 400 aviation and aerospace manufacturers from 25 countries and regions. 96% planes from around the world were exhibited and gave performances. Most striking (to Chinese audiences, anyway) among the exhibits were over 50 aerospace products and models, including carrier rockets, defense missiles and man-made earth satellites provided by the China National Aerospace Industrial Corporation. During the exhibition, a symposium on "Chinese Aviation in the 21st Century" aviation and aerospace exchange meetings and trade negotiations were held. The 6-day exhibition also saw an influx of 800,000 spectators from all parts of the world, filling up overnight all the hotels in Zhuhai.

S: What about the opening ceremony of the ZIAAE?

E: On the morning of Nov. 5, 1996, the opening ceremony of the airshow was held in Zhuhai, attracting over 7,000 guests, including state, provincial and municipal leaders. China's Premier Li Peng cut the ribbon and Wu Bangguo, Vice Premier of the State Council and

Director of the '96 China International Aviation and Aerospace Exhibition Organizing Committee, delivered the opening address. Then, women parachutists from the Parachute Team under the State Physical Culture and Sports Commission gave a wonderful jump, raising the curtain of a stunt flight performance.

S: I guess that the flight performances of various planes must have been very wonderful and exciting.

E: Definitely. The first performer was the F-8IIM, "China's master of the skies" and the pride of the Chinese Air Force. The graceful and stream-lined fighter was seen whistling through the air, executing various maneuvers, somersaulting, turning round, and then doing a loop. Smoothly linking one demanding movement after another, the fighter impressed many spectators. Soon after the F-8IIM touched ground on its landing approach, a K-8 trainer ascended into the sky, and drew the spectators' attention by rolling horizontally. Now came the turn of the Z9A helicopter. It first hovered motionless for a while, then dived towards the ground almost perpendicularly, and before reaching the ground ascended again in a straight line. After these rousing performances the F-7 also gave a wonderful performance. In recent years China has manufactured the more advanced fighters such as F-III, J10 etc. These breath-taking performances were followed by stunt flights performed by the famous Golden Dream Acrobatic Team from Britain, and the Su-27 and Su-30 from Russia.

S: Were there many models of various rockets and satellites at the airshow?

E: No doubt. After the flights, 2,000 rockets on the exhibition were all launched. On the west end of the exhibition ground stood the Long March 2 Cluster Carrier Rocket, a telling witness of China's achievements in aerospace. It is one of China's many types of Long March carrier rockets and can launch man-made satellites into orbit, including a perigee orbit, solar orbit and earth static orbit, and send a 9-ton satellite into perigee orbit and 5-ton satellite into an orbit 36,000 kilometers from earth. To date, the Long March rockets, together with other "storm rockets" have launched 47 satellites after 42 launches, 10 of which were for foreign customers. International insurance firms claim that the rate of successful launches for China's Long March rockets is 85%.

S: How about the strategic and tactical missiles of China?

E: China already possesses powerful liquid and solid strategic missiles, an efficient anti-nuclear force and various integrated tactical missiles; has successfully developed ground-to-air missiles, air defense missiles; winged missiles including ground-to-ship, air-to-ship and ship-to-ship missiles; and has also succeeded in developing a supersonic anti-ship series and marine defense series.

S: Has China also attained great achievements regarding the recovery of satellites?

E: Sure. On Oct. 20, 1995, China launched a satellite for scientific exploration and technological tests, which, after 15 days in its orbit, was successfully recovered on Nov. 4, 1995, in central Sichuan Province. Since it launched its first recoverable satellite in 1975, China has launched 17 such satellites and recovered 16 of them successfully. The remote sensing materials they brought back and the achievements of the scientific experiments conducted in outer space have tremendously benefited the nation's economy and social development. In all, China's aerospace industry has established a sound base for the defense of the country.

S: What happened behind the airshow?

E: The AE-100, the world's first civilian plane program primarily for Asia, is being co-sponsored by China, Singapore and the European firms of Aerospatiale, British Aerospace and Alenia of Italy. With China holding a 49% share, the European companies 39% and Singapore 15%. In order to explore China's civilian aircraft market, the Russian delegation not only brought with them the Su-27 and Su-30, but also demonstrated the civilian Tu-204. The Russians quickly signed a contract with China for 20 Tu-204s. Meanwhile, Boeing, Mc-Donnell Douglas, Aerospatiale, British Aerospace, and Russian Avia export were also eager to use this opportunity to display their aircraft and attract buyers. The Zhuhai Airshow concluded with 15 projects involving US＄2 billion. It is estimated that China will need 1,350 more planes in the next 20 years. The 96th China International Aviation & Aerospace Exhibition drew to a successful end on Nov. 10, leaving pleasant memories to all present at the show.

Notes

1. Zhuhai International Aviation and Aerospace Exhibition (ZIAAE) 珠海国际航空航天展
2. William E. Boing 威廉 E. 波音
3. Conarad Westervelt 康拉德·韦斯特维尔特
4. Dashatou Aviation Bureau in Guangdong 广东大沙头航空局
5. Dr. Sun Yat-sen 孙中山先生
6. photographed 拍了照片的，被摄影出的
7. Madame Soong Ching Ling 宋庆龄女士
8. great fanfare 巨大轰动
9. Aviation Industry 航空工业
10. make aircrafts on license 根据许可证制造飞机
11. launching man-made satellites 发射人造卫星
12. fighters, bombers, attack planes, helicopters, air ferries and civilian passenger planes 战斗机、轰炸机、攻击机、直升机、运输机和民航客机
13. multiplied 大大增加
14. the International Civil Aviation Organization 国际民用航空组织
15. at a lightning speed 以闪电般速度
16. turn the country's long-cherished dream into reality 将这个国家长期的梦想变为现实
17. carrier rockets, defense missiles 运载火箭，防御导弹
18. the China National Aerospace Industrial Corporation 中国国家航天工业公司

19. saw an influx of 800,000 spectators 参观人流量达 80 万
20. cut the ribbon 剪彩
21. delivered the opening address 致开幕词
22. Women parachutists from the Parachute Team under the State Physical Culture and Sports Commission gave a wonderful jump, raising the curtain of a stunt flight performance. 隶属国家体育文化和运动委员会跳伞队的女跳伞运动员揭开了特技飞行表演的序幕。
23. China's master of the skies 中国蓝天的主人
24. the graceful and steam-lined fighter 优美的流线型战斗机
25. whistling through the air 呼啸升空
26. executing various maneuvers, somersaulting, turning round and doing a loop 表演各种灵巧的动作，向前、向后翻筋斗，转身，然后又翻筋斗
27. smoothly linking one demanding movement after another 一个个规定动作平稳而连贯
28. touched ground on its landing approach 在其跑道着陆
29. a K-8 trainer ascended into the sky K-8 教练机升空
30. by rolling horizontally 水平翻滚
31. hovered motionless for a while 平静地在空中盘旋一会儿
32. then dived towards the ground almost perpendicularly and before reaching the ground ascended again in a straight line 几乎呈垂直向地面俯冲，在触及地面之前又垂直升空
33. rousing performances 令人振奋的表演
34. by stunt flights performed by the famous Golden Dream Acrobatic Team from Britain 著名的英国金梦杂技队表演的特技飞行
35. Su-27 and Su-30 from Russia 俄罗斯的苏-27 和苏-30 型战斗机
36. the Long March 2 Cluster Carrier Rocket, a telling witness of China achievements in aerospace 长征二型捆绑式运载火箭是中国航天成就的有力见证
37. a perigee orbit, solar orbit and earth static orbit 近地点轨道，太阳轨道和地球静态轨道
38. storm rockets 风暴火箭
39. launched 47 satellites after 42 launches 42 次发射后发射了 47 颗卫星
40. possesses powerful liquid and solid strategic missiles, an efficient anti-nuclear force and various integrated tactical missiles 具有以液态和固态为强大动力的战略导弹，有效的反核力量和各种综合的战术导弹
41. ground-to-air missiles, air defense missiles; winged missiles including ground-to-ship, air-to-ship and ship-to-ship missiles 地对空导弹，空中防御导弹；包括地对舰、空对舰、舰对舰的飞翼导弹
42. a supersonic anti-ship series and marine defense series 超声反舰系列和海上防御系列导弹
43. the recovery of satellites 回收卫星
44. first recoverable of satellites 第一颗可回收卫星
45. the remote sensing materials that they brought back 他们带回的遥感材料
46. the achievements of the scientific experiments conducted in outer space 在外太空进行科学实

验的成就
47. the AE-100 亚欧-100 型民航飞机
48. co-sponsored 共同主办（发起）
49. Aerospatiale, British Aerospace and Alenia of Italy 帕蒂阿尔航空公司、英国航天公司和意大利阿利尼亚航空公司
50. the civilian Tu-204 图-204 民航飞机
51. Mc-Donnell Douglas 麦克-唐纳道格拉斯公司
52. Russian Avia export 俄罗斯航空出口公司
53. concluded with 15 projects involving ＄2 billion 达成了 15 个工程项目，涉及金额达 20 亿美元
54. drew to a successful end 取得圆满成功

22. Innovations in Business
（企业创新）

B: You're a specialist in innovative business techniques, Mr. A, would you explain what is innovation in business, especially from the standpoint of management?

A: Well, the concept of innovation really means working out a new idea and then developing and implementing it.

B: Is it an essential part of all business management?

A: Yes, it is. Actually I would say that innovations are the real life-blood of business management. As you know, we need growth and development to survive. Some people say that if we don't grow, we die.

B: Do you think the same is true for businesses?

A: Sure. Businesses really cease when they run out of ideas. And managers must put these new ideas into practice where they can be effective and mean something to the firm.

B: So it is a part of the dynamics of the management cycle.

A: Exactly, it must be a continuous process of dealing with and implementing new ideas.

B: But we often find it difficult to decide whether a company should develop new technology itself or purchase technological development on the outside in terms of innovation.

A: A small firm faces that problem more often. We keep our own research development section, but we also license technology from other companies.

B: Yes, we keep the same practice. In several cases, it has been a cross-licensing agreement with innovations we have developed ourselves.

A: It is necessary, the know-how market has become a major aspect of the modern manufacturing process. In many cases, we have been able to take advantage of significant technological

innovations for a small consideration.

B: Several times we purchase the use of a patent right on a royalty basis. Of course, this situation usually applies in product patents.

A: Yes, process patents or manufacturing technology are usually licensed by initial payment and subsequent installments based on output.

B: Right, licensing arrangements is likely to keep a firm competitiveness.

A: In addition, information is essential to the achievement of more efficient and effective management.

B: Yes, that's why computers have been very valuable in collecting, codifying and analyzing information quickly and accurately. Some computerized data systems like E. D. P., or Electronic Data Processing, and I. D. P., or Integrated Data Processing really help a lot in decision making.

A: That's true. And there is a more recent M. I. S., or Management Information System, which can gather the valuable raw data very quickly and accurately.

B: So we can also say taking the advantage of computerized data system is an innovation in management technique.

Notes

1. innovation in business 工商企业的创新
2. specialist 专家
3. implementing 实现；落实
4. life-blood 生命线；命根子
5. a part of the dynamics of the management cycle 管理循环中有活力的部分
6. research development section 研究开发部门
7. license technology from other company 从其他公司获得许可证技术
8. cross-licensing agreement 互签的许可证协议
9. know-how market 专门技术市场
10. In many cases, we have been able to take advantage of significant technological innovations for a small consideration. 在许多情况下，我们已经能够利用重大的技术创新成果，而只偿付较少的报酬。
11. patent right on a royalty basis 在专利使用费基础上的专利权
12. Process patents or manufacturing technology are usually licensed by initial payment and subsequent installments based on output. 加工技术专利或制造技术许可证的获得通常需要交付首期付款加上以后按产量计算的分期付款。
13. licensing arrangements 许可证协议
14. codify 编撰；整理

15. computerized data system 计算机数据系统
16. Electronic Data Processing（E. D. P.）电子数据处理系统
17. Integrated Data Processing（I. D. P.）集成数据处理系统
18. Management Information System 管理信息系统
19. raw data 原始数据

23. China International High-tech Exhibition
（中国国际高新技术展览会）

A：China International High-tech Exhibition is held by the China Council for the Promotion of International Trade（CCPIT）every autumn in Beijing. It is our great pleasure to inform you that the exhibition will display a great variety of the most advanced and up-to-date technologies and products, including IT, manufacturing facilities, biological engineering, digital technology, high efficient ecological agriculture, aerospace industry, robots, intelligent（smart）techniques, automobiles and so on. We are sure that this exhibition will offer you some extremely favorable opportunities to push the sale of your products in China and the world. As you have made tremendous efforts in the past to sell our new products, we believe you will take advantage of the chance to undertake sales promotions so as to lay a solid foundation for promoting sale both during and after the exhibition, and let us know if there is anything we can help you.

B：Thank you very much for your information. Our corporation will surely take part in the exhibition held by the CCPIT. International fairs and exhibitions play an important role in broadening economic relations. Would you please tell us something about the exhibition area and the companies that will participate in the exhibition?

A：Sure, this will be the first large scale international fair and exhibition ever organized in China. More than 50 countries and regions will participate in the exhibition. It involves nearly 300 companies, among which two fifths of the corporations will come from the *Fortune* 500 companies. Some developing countries and regions will exhibit their exports that have great potentials. The total exhibition area covers 60,000 square meters of indoor floor space and 25,000 square meters of outdoor exhibition area. The Chinese pavilion occupies over 10,000 square meters. The pavilions are designed in traditional Chinese style. They can be divided into several exhibition rooms.

B：Well, our corporation will bring lots of new products, such as digital TV sets and digital cameras etc. to the exhibition for promotion. We want to rent two or three booths. The area of each booth should not be less than 150 square meters. What is the least expensive booth you have?

A: We have some spaces left in the middle and in the back of the Exhibition Hall, which may meet your requirements. The rent will be $1,200 each per day in the middle and $1,000 each day in the back respectively.

B: OK, then we book two booths in the middle of the Exhibition Hall. In order to promote the sale of our products at your markets, we are projecting to concentrate our TV publicity on your commercial television stations at golden time one week before and during the exhibition. We also wish to appoint representatives in the most important cities in your country and intend to advertise in several newspapers. We hope that both our exhibits and products can be disposed of and sold locally after the fair comes to end. Would you give us your advice as to which papers would be the most suitable for this purpose?

A: Yes, our TV advertisement has unparallel excellence on CCTV and BJTV despite their expensiveness. We agree that it will be best to insert advertisement in the leading newspapers of our most important cities and we shall fax a list of the most suitable newspapers, TV stations together with their charges for you to choose.

B: This is a good idea. Perhaps we shall ask you as our agent. At the exhibition, we will make negotiations with the Chinese exporters and importers on goods China needs and on commodities American demands.

A: The exhibition is aimed at increasing mutual understanding between China and other countries and regions, and at promoting trade and scientific and technical cooperation between them. The exhibition will display Chinese export commodities for foreign business people to make selections in purchasing against samples in business talks on the spot. And furthermore, the stand-attendants can show you the operation of the most equipments. I am confident that the novel designs of exhibits at the International High-tech Exhibition will attract hundreds upon thousands of entrepreneurs, engineers, technicians and businessmen. Business activities and talks will be brisk, and the deals estimated at the exhibition will exceed three and a half billion US dollars.

B: I think it will be promising! Thanks a lot.

A: I am looking forward to meeting you at the exhibition!

Notes

1. China International High-tech Exhibition 中国国际高新技术展览会
2. highly efficient ecological agriculture 高效生态农业
3. intelligent (smart) techniques 智能技术
4. the *Fortune* 500 companies 《财富》500强公司
5. the Chinese pavilion 中国展区
6. booths 摊位、售货棚
7. disposed of 处理、卖掉

8. to make selections in purchasing against samples in business talks on the spot 使他们在现场的商务洽谈中能根据样品选购货物
9. The stand-attendants can show you the operation of the most equipments. 讲解员会给你们进行大多数设备的操作演示。
10. the novel designs of exhibits 展品的新颖设计
11. brisk 踊跃，活跃，繁荣，生气勃勃

Part Two ❷

Technology Transfer

技术转让

24. Technology Transfer (1)
(技术转让之一)

Mr. Ma, a senior engineer (E) and a student (S) who is studying international trade are discussing technology transfer.

S: Mr. Ma, I have a question to ask.

E: Go ahead!

S: What's the implication of technology?

E: Technology mainly consists of systematic scientific knowledge, ripe practical experience and operating techniques. It is a special expertise and has been gradually formed in the long activities of production on the basis of combination of the principles of natural science and experience of social practices. Technology is directly utilized in the whole process—from study and development of the products to sales of them. It may not only take inventions, utility models, industrial designs or new breeds of plants and animals as its forms of expression, but also find expression in technology information, technology specifications, technical standards, techniques and skills in different kinds of services, such as installation, establishment, operation, debug, maintenance, management of commerce and enterprises, and in processing raw materials and equipment.

S: How many kinds does technology usually fall into?

E: Generally speaking, technology falls into three kinds: ① Open Technology. This technology commonly includes scientific and technical theories and practical knowledge that do not belong to industrial property, and can often be passed on or enjoyed the wide use freely. Open technology is known to the public, therefore, the contents of which may be found in different scientific and technological books, writings, magazines, thesis, reports and even data of academic seminars, symposiums etc. ② Semi-overt Technology. It is usually referred to as patent technology. Semi-overt technology consists of rights of patent which are obtained through patent applications and grants of them. Within the patent limitation, patent technology falls into the scope of industrial property and can't be copied or used without permission. According to patent laws, all the inventions must be made open fully. But actually, they are often kept in a semi-overt state, the key technology is still kept confidential. ③ Secret Technology. Secret technology is commonly referred to as know-how and not known to the public. In most cases, it is usually a kind of key technology which is not protected by any special laws. In order to be kept secret in technology transfer, only the laws made to protect commercial secrets may be invoked. However, with the development of science and technology, patent technology and know-how will be changed to open technology in the process of propagation and transfer. In technology trade, either acquisition or transfer of technology, mainly focuses on patent technology and trade marks, which are protected by special laws of industrial property, and know-how, which is not protected by any special laws of industrial property.

S: What's the purpose of acquisition of technology for a country or a corporation?

E: Acquisition of technology is the most important form of commercial trade in many countries. Our country mainly imports foreign advanced technologies in equipment or in management. It is of great benefit to us not only in the short run, but also in the long run. The major purposes may be summarized into four: ① Introduction of foreign advanced technology will help promote the development of the national economy and make it come up to a higher level within a shorter period of time. ② Technology importation will be advantageous to the industrial production and manufacture of equipment of a country. By making use of existing technology and design techniques, people may quickly make their industry upgrade and catch up with the advanced world level, and lay a reliable foundation for further development of a country's national industry in the course of assimilating the introduced technologies and improving them. In this way, one country will certainly save much investment, and much time as well. ③ Acquisition of technology will greatly promote the improvement of industrial productivity and the quality of products of a country. Only in this way can a state continuously expand the sales of its products in the world market and further strengthen the competitive power in international trade. ④ By assimilating foreign advanced technology, one country may much easier bring up a great number of qualified scientists and technicians, and experts of modernized business management, and much more quickly raise the level of science and technology.

S: Would you please give us a brief explanation of technology trade?

E: OK! Technology trade developed in the world after World War II. Since the 1960s it has been becoming one of the most important forms in international trade. Especially in recent years, technology trade has made rapid development in the world. It has been proved that both the developed and developing countries have benefited from technology trade. According to the United Nations, the total amount of technology trade was about 3 billion U.S. dollars in 1965; 11 billion in 1975; 50 billion in 1985, and is estimated to reach 260 billion dollars in 1995.

S: What are the major features of technology trade in recent years?

E: ① At present, the international technology trade is chiefly dominated by five developed countries and more than 70% of the total amount is shared by the U.S.A., England, France, Japan and Germany. About 10% of that is obtained by other developed countries in the West. The amount of technology trade in the former Soviet Union and Eastern European Countries comes to another 10%. That of the developing countries only accounts for the remaining 10%.
② Relying on the technological superiority, the developed countries try their best to export technology to other countries, especially to the developing countries so as to seek great profit. According to statistics, they will sell out 5 billion U.S. dollars full sets of equipment and other products when they transfer 100 million U.S. dollars patent technology by stipulating the tie-in

clause in contracts. On the other hand, about 80% of technology trade is occupied by the developed countries, the competition is getting more and more heated, particularly in the trade war of advanced and up-to-date technologies.

S: Are there any more?

E: Yes. ③ In the international technology trade, the transnational companies always find themselves in a monopolistic position. Since the 1980s, transnational corporations have been developing rapidly. At present, the developed countries make most of their foreign investment, that is to say, about 80%, through these big corporations. In practice, capital exportation is closely combined with technology exportation. According to statistics, 90% technology trade in the West is monopolized and controlled by 500 biggest transnational companies in the world. ④ Almost all the developed countries have already changed their strategies in capital exportation. They make full use of technology exportation to bring along capital and commodity exportation. At present, they do, by every possible means, export technology of lower levels or even technology which is about to fall into disuse to developing countries. In this way, they can not only earn plentiful fees of technology transfer, but also promote equipment and capital exportation. ⑤ Currently, a new situation of multilateral technology trade has formed. The technology trade relations have been developing not only between the developed and the developing countries, but also between the developed countries themselves and between the developing countries themselves.

Notes

1. ripe practical experience 成熟的实践经验
2. a special expertise 一种特殊的专门知识
3. It may not only take inventions, utility models, industrial designs or new breeds of plants and animals as its forms of expression. 它不仅可以用发明、实用模型、工业设计、植物的新种子、动物的新品种作为其表现形式。
4. debug 排除故障，调试
5. can often be passed on or enjoyed the wide use freely 通常能够流传下来或者可以自由地享有广泛的使用权
6. Semi-overt Technology 半公开技术
7. through patent applications and grants of them 通过专利申请并获得专利权（许可）
8. confidential 机密的、秘密的
9. know-how 专门技术
10. only the laws made to protect commercial secrets may be invoked 只有为保护商业秘密而制定的法律才可以行使法权（才可以付诸实施）
11. propagation 传播、宣传
12. acquisition 购买、获得

13. assimilating 吸收消化
14. by stipulating the tie-in clause in contracts 在合同中规定了连带条款
15. advanced and up-to-date technologies 高新技术
16. monopolistic position 垄断地位

25. Technology Transfer（2）
（技术转让之二）

S：What are the ways of international technology transfer?

E：Technology transfer refers to the transfer of systematic knowledge by one party to another in producing a certain product, applying a certain technique or skill, or in providing a certain service to the latter, but not including only sales of goods or lease business. The transfer does not involve the assignment of technology ownership but the right to use technology. Many ways of international technology transfer may be taken.

S：Are joint ventures a form of technology transfer?

E：No doubt. Joint ventures are a form of business relations which involves varying degrees, pooling of assets, joint management and a sharing of profits and risks according to a commonly-agreed formula. They may be made up of two or more private businesses, enterprises, companies, or other economic organizations, or may be even by governmental departments or international institutions. They jointly invest in and together manage the same business. There are two basic types now in operation in China：① contractual (non-equity) joint venture; ②equity joint venture.

S：What about project contracting?

E：It is a way of technology transfer. Project contracting refers to a contract or an agreement for construction of a civil engineering project concluded by the contractor with the project owner. These projects are often big items of new construction. It will take a great sum of fund and a long period of time for the contractor to complete the whole construction work. Therefore, it is rather difficult to undertake such a task and carry out the contract without any risks. Very often, the force majeure, price adjustment, hardship and other special clauses are expressly and properly stipulated in the contract, so as to lessen the risk borne by the contractor. These contracts or agreements of engineering project include：①turn-key contract; ②product-in-hands contract; ③ semi turn-key engineering contract; ④ contract for civil engineering; ⑤contract for project construction; ⑥contract for supply and installation of equipment.

S：I think equipment importation is another kind of technology transfer. Am I right?

E：Quite right. By the way of purchase contracts, we may import various kinds of technical equipment urgently needed to develop our own national economy and heighten the level of production technology. The importation of equipment is related to the acquisition of technology, and it should include the transfer as to license of industrial property, license of know-how or technical service, otherwise no difference can be made between equipment importation and common purchase of products. Equipment importation usually includes: ① full set of equipment; ② whole production line; ③ turn-key project; and ④ key technical equipment. In China, the transaction is usually a complete package containing the equipment, the right to use industrial property or know-how related to the item, and other technical knowledge and techniques needed in installation, operation, or maintenance.

S: Is cooperative production a way of technology transfer?

E: Of course! This arrangement is also called technical collaboration, or shortened to co-production, under which one party undertakes to cooperate with the technology supplier to manufacture a certain type of industrial product in accordance with the engineering instructions, industrial property or know-how supplied by the later. Sometimes according to the agreement, both parties manufacture different parts of the same products, and then they are assembled into finished products by one party or both. In China, the objective of cooperative production centers on our acquisition of necessary technology from abroad to produce the products all by ourselves. So the two basic forms of technical collaboration are commonly used by Chinese enterprises.

S: What are they?

E: ① An agreement of cooperative production is usually concluded between the two parties, whereby the technology supplier shall provide the technical instructions, blue-prints of design, or even equipment for production, so that the party in China is able to start manufacturing parts and components or certain products for export or domestic sales in accordance with the stipulations in the agreement. ② Sometimes the technology supplier shall be solely responsible for the industrial property and know-how, and cooperate with a Chinese enterprise according to the terms and conditions laid down in the contract.

Notes

1. assignment of technology ownership 技术所有权的转让
2. pooling of assets 资产联营
3. a sharing of profits and risks according to a commonly-agreed formula 根据双方同意的准则共享利润，共担风险
4. project contracting 工程承包
5. the force majeure 不可抗力
6. turn-key contract 总承包工程（交钥匙工程）合同
7. product-in-hands contract 产品到手工程承包合同
8. the transaction is usually a complete package 该交易通常是完整的一揽子的（包干的）交易

9. arrangement 协议
10. technical collaboration 技术合作
11. co-production 共同生产

26. Technology Transfer (3)
（技术转让之三）

S: Is license trade a broadly used technology transfer in the world?

E: Sure. By signing a contract or an agreement, one party transfers industrial property or know-how to another party. The latter will enjoy the right to use the technology, including rights of manufacture and sales, but not the right of technology ownership. In license trade the licenser will supply technology, and the licensee will pay fees for using the technology. License trade, therefore, is to transfer the right to use industrial property or know-how, not the ownership of industrial property or know-how itself. A licensing contract or agreement will be arrived at between the technology supplier and the technology acceptor. The right granted by the licenser to the licensee, may refer to patents, trade marks, or know-how, or those of the mixed two or three, and may be granted through exclusive license, nonexclusive license, sole license, sublicense, or cross-license. In transfer, the supplier is usually obliged to furnish relative technical information and assistance, and the licensee will have to make use of the right in accordance with the stipulations in the agreement and pay the agreed amount of fees to the former.

S: Are the types of license contracts mentioned above now commonly used in technology transfer?

E: Yes. ① Patent license. In a patent license contract, the patent is taken as the contracted subject matter. Under this contract, the licensee is allowed to make use of the patented technology and pays to the licensor the agreed amount of fees for using the technology. ② Trade mark license. A contract entered into both parties is called a trademark license contract. The registrant allows the other party to use the trademark within the stipulated territory and time limit, and the acceptor pays the agreed amount of fees to the owner of the registered trademark. ③ Know-how license. When a know-how license contract is reached, the supplier shall grant the right to the acceptor to make use of the secret technology, and at the same time the acceptor is responsible for keeping secret in accordance with the confidential clause and pays the fees for using the know-how. ④ Mixed license. Sometimes, if the subject matters are at least two or three items, the contract concluded by the both parties will be named as a mixed license contract. In license trade, it becomes the chief form because the transfer of right of both the patent or trademark and know-how is the most common business in technology trade.

S: According to different rights transferred, how many types of rights do we usually have?

E: There are five. ① Exclusive license. When this type is adopted, the licensee will have the right to use exclusively the technology transferred to produce and sell the licensed goods in the territory designated in the contract or agreement. The licensor himself only has the right to sell the said products in other areas than the specified territory. Once a contract is entered into, the supplier shall not be allowed to transfer the technology to any third party at his discretion. ② Sole license. The sole license is different from the exclusive license in the designated territory. The licensee has the privilege of exclusive use of the transferred technology to produce and sell the licensed goods, in the agreed region. But according to the terms and conditions stipulated in the sole license contract the licensor himself retains the right to sell the same licensed goods in this area. The licenser shall confine himself not to granting any license to any third party in this territory. ③ Nonexclusive license (simple license). When a nonexclusive license contract is concluded, both parties shall agree that both the licensor and the licensee have the right to produce and sell the licensed products in the designated territory, and at the same time the licenser may transfer the said technology to a third party in this area. Because the fees for using technology are much lower, this type is popularly used in China and some developing countries.

S: Then the other two?

E: Well. ④ Sub-license (retransferring license). Since the licenser agrees that the licensee may transfer the right to use the said technology to a third party in the designated territory on condition that he pays more fees to the supplier after he has got the right to produce and sell the licensed products in that area. A sublicense contract is usually subordinate to a nonexclusive license contract. ⑤ Cross-license. It is a special type in international license contracts or license agreements. In the contract, the parties concerned stipulate that both sides shall exchange the rights to use the technologies to produce and sell the licensed products with each other, which are approximately of equal value, on the basis of equality and mutual benefit. This type only appears in cooperative productions and designs or joint ventures. Usually no payment will be needed and the rights transferred may or may not be exclusive.

Notes

1. through exclusive license, nonexclusive license, sole license, sublicense, or cross-license 通过独占许可、普通许可、排他许可、从属许可、互换许可
2. the contracted subject matter 合同标的物
3. registrant 商标注册人
4. at his discretion 随意、任意
5. subordinate to 从属于、次要的

27. Technology Transfer (4)
(技术转让之四)

S: What about franchising?

E: By concluding a franchising contract or agreement, an industrial or commercial enterprise that possesses rich and successful experience, grants the right to use its trademark, name of firm, service mark, patent, know-how, methods and experience in management to another enterprise. Both parties will adopt the same modes of business management: ① engage in the same business line; ② sell the same commodities; ③ provide the same services in the same manner; ④ use the same names of commodities, trade marks, or service marks; ⑤ decorate their shops or stores with the same materials, in the same ways; ⑥ furnish the rooms or halls, with the same furniture; ⑦ make the products, by using the same methods; ⑧ wear the same kind of working clothes. However, the two enterprises will operate independently and are responsible for either their own profits or losses respectively.

S: Is compensation trade a very popular mode of trade both in China and some developing countries?

E: Undoubtedly, in compensation trade an overseas party shall provide a Chinese enterprise with advanced equipment and machinery, technical service, and even some necessary raw materials, and the latter shall be responsible for processing or producing certain products, and make compensation for the former's equipment and relative technology within the time limit agreed by the two parties in the contract. According to different products compensated, compensation trade can be classified into two types. ① product Buyback: The products compensated for the equipment, technology and raw materials are directly manufactured by the said equipment and technology. ② counter Purchase (indirect compensation): The said equipment and technology are compensated by other commodities or services agreed by the two parties in advance.

S: Is consulting service a type of technology transfer in international technology trade?

E: Certainly. Consulting service refers to the technology supply in services, through which one may apply one's various kinds of scientific and technical knowledge, practical experience or skills, in the form of technical data, information, survey reports or proposals to the technological items designated by the client. This service is offered to acquaint the client with the feasibility of a certain project or to train the client's personnel to master and use the said technology or expertise. In practice, technical consulting service may take different forms to suit different situations: ① Feasibility Study of Engineering Projects; ② Programs of Technology Designs; ③ Appraisal of Production, Quality control, and Marketing Programs;

④ Investment Plans and Scenario; ⑤ Regional Development Plans; ⑥ Technical Assistances and Instructions in Equipment Assembly, Installation, Debug, and Maintenance; ⑦ Technical Training Service; ⑧ Business Management Schemes.

S: Mr. Ma, you haven't mentioned cooperative exploitation. Is it a mode of technology transfer?

E: Yes, it is. Now in China as well as in some developing countries, this type is usually adopted in exploiting oil fields, mines and other natural resources.

S: What are the main clauses in the contract of technology acquisition?

E: The contents of the contract, namely the provisions, are normally agreed by both parties to govern the rights and liabilities stipulated in the contract. There are many different forms of contracts available for proper circumstances, but the following main clauses are basically included:

1. Preamble
①Title of Contract; ② Signing Parties; ③ Each Party's Authority; ④ Place of Signing; ⑤Date of Signing; ⑥Whereas Clause or Recitals; ⑦Definition Clause; ⑧Third Party.

2. Terms and Conditions
① Basic terms and conditions, including technology scope, technology characteristics and technology documentation.
② Correct selection of technology, including limitation of technology, supporting technical data, national standard system.
③ Reasonable conversion of technology documentation.
④ Exchange of improved technology, including ownership, rights and duties in exchange, methods used to handle great improvement of technology.

3. Right Granted
① Right to use technology. ② Right to manufacture the licensed products;
③ Right to sell the products. ④ Marketing territory.

4. Price and Payment
① Evaluation factors, including effective value, market shares, technology level, technology sources, contract conditions.
②Terms of price, including lump-sum payment, royalty (fixed royalty, sliding royalty, minimum royalty and maximum royalty), initial fees or disclosure fees and royalty.
③ Payment, including payment instrument, payment method, and payment terms.

5. Trade Mark
① Indication of trademark. ② Combined trademark. ③ Associated trade mark.

6. Technical Training and Technical Service
① Technical training. ② Technical service.

7. Technology Acceptance
① Acceptance of technology documentation. ② Acceptance of products.

8. Confidential

① Confidential scope. ② Confidential area. ③ Confidential limitation of time.

9. Warranty (or guarantee) and Claims

① Difference between warranty and guarantee.
② Warranty (or guarantee) clause, including warranty (or guarantee) for technology documentations, product properties and results, rights granted.
③ Remedial measures taken against breach of contract and methods used to make settlements, amount of penalty or compensation, guarantee letter of the bank.

10. Taxes

① Double taxation. ② Single taxation. ③ Foreign tax credit.

11. Disputes and arbitration

① Place of arbitration. ② Arbitration institution. ③ Arbitration methods and rules. ④ Arbitration procedures. ⑤ Effect of award. ⑥ Arbitration fees.

12. Force Majeure

① Accident. ② Legal consequence.

13. Duration, Termination, Assignment, Amendment of Contract and Governing laws

① Duration of contract. ② Termination of contract. ③ Assignment of contract. ④ Amendment of contract. ⑤ Governing laws of contract.

14. Witness Clause

① Concluding contents. ② Signature. ③ Seal.

S: Well, you gave me a detailed explanation of technology transfer, and I have learned a lot about it from you. Thank you very much.

Notes

1. service mark 服务标识、服务商标
2. engage in the same business line 从事相同的业务
3. product buyback 产品返销
4. counter purchase 回购或互购
5. to acquaint the client with the feasibility of a certain project 使客户了解某一工程的可行性；通知客户某一工程的可行性
6. scenario 远景预测，方案
7. preamble 导言，序文
8. Each Party's Authority 当事人合法依据
9. Whereas Clause or Recitals 约由（条款或说明）
10. Definition Clause 定义条款
11. supporting technical data 配套技术资料
12. reasonable conversion of technology documentation 技术文件的合理更换

13. lump-sum payment 一次性全额支付
14. royalty (fixed royalty, sliding royalty, minimum royalty and maximum royalty) 提成支付（固定提成费、浮动提成费、最低提成费、最高提成费）
15. payment instrument 支付票据，支付工具
16. combined trademark 联合商标
17. associated trademark 联合商标
18. remedial measures 补偿措施，赔偿措施
19. breach of contract 违约
20. amount of penalty or compensation 罚金或补偿金额
21. guarantee letter of the bank 银行担保书
22. foreign tax credit 国外税款抵免
23. effect of award 裁决效力
24. duration, termination, assignment, amendment of contract and governing laws 合同的期限、终止时间、转让、修改与适用法律
25. Witness Clause 结尾条款

28. The Royalty Rate and the Initial Down Payment Are Too High （提成费和入门费太高）

Mr. Zhang, representing Sinopec (Shanghai Branch) needs high technology to upgrade the factory, and Mr. White, representing London Trade Company is willing to provide technological assistance in the form of license. They are still negotiating.

Zhang: Shall we take up business now?

White: That's fine with me. I'm ready.

Zhang: We need high technology to renovate our factory.

White: I hope we can satisfy your requirements. Our company has the most advanced technology for both production and management in the world. We can supply all licenses, equipment, engineering service and technical assistance.

Zhang: Very good. Your license should guarantee that machines and technology are of advanced world level and the technology provided is integrated, precise and reliable.

White: No doubt about it. Here is an introduction booklet of our products and production technology.

Zhang: The introduction is really detailed. What right will the license grant to our factory?

White: It will grant rights of both manufacture and sales of the products.

Zhang: Does the license include the patent?

White: Yes and the validity of the patent is 15 years from now on.

Zhang: What about the payment for the technology?

White: Which would you prefer, lump-sum payment or a combination of an initial payment and a royalty?

Zhang: We prefer the combination of initial payment and royalty.

White: Then in addition to an initial down payment of 200,000 US dollars the royalty is 5% of the net sales price of the products.

Zhang: I'm afraid both the royalty rate and the initial down payment are too high.

White: I'm surprised to hear that. You know, the sum can hardly cover all our expenses in technological information, drawings, personnel training and so on.

Zhang: We propose that you lower the royalty rate to 3% of the net sales price.

White: Well, 3%... it's OK.

Zhang: In that case we accept the amount of 200,000 US dollars as the initial down payment.

White: We have reached an agreement on payment because of our sincere desire. Then, within 30 days after the effective date of the contract, you must pay the initial payment of our license. Besides, you must pay us a running royalty of 3% of the net sales of the licensed products during the period of 10 years after the effectuation of the contract.

Zhang: That's acceptable.

Notes

1. The royalty rate and the Initial down payment are too high. 提成费和入门费太高。
2. Sinopec (Shanghai Branch) needs high technology to upgrade the factory 中国石化公司（上海分公司）需要高新技术以将工厂升级
3. provide technological assistance in the form of license 愿意以许可证方式提供援助
4. Shall we take up business now? 我们现在谈生意（业务）好吗？
5. renovate our factory 改造我们的工厂
6. the most advanced technology 最先进的技术
7. We can supply all licenses, equipment, engineering service and technical assistance. 我们能够提供各种许可证、设备、工程服务和技术援助。
8. Machines and technology are of advanced world level and the technology provided is integrated, precise and reliable. 机器和技术具有世界先进水平并且所提供的技术是完整的、精确的和可靠的。
9. Here is an introduction booklet of our products and production technology. 这是一本介绍我们产品和生产技术的小册子。
10. What right will the license grant to our factory? 该许可证将会给予我们工厂什么权利？
11. Does the license include the patent? 该许可证包括专利吗？
12. the validity of the patent 专利的有效期

13. What about the payment for the technology? 这项技术如何支付款项？
14. Which would you prefer, lump-sum payment or a combination of an initial payment and a royalty? 你们愿意一次性付清还是采取入门费加提成费方式付款？
15. Then in addition to an initial down payment of 200,000 US dollars the royalty is 5% of the net sales price of the products. 那么，除了20万美元的首期付款以外，许可证的提成费比例是其产品净销售额的5%。
16. The sum can hardly cover all our expenses in technological information, drawings, personnel training and so on. 这笔钱难以支付技术信息、图纸和人员培训的花费。
17. We propose that you lower the royalty rate to 3% of the net sales price. 我们提议你将提成费降低到其净销售额的3%。
18. because of our sincere desire 因为我们双方的诚意
19. within 30 days after the effective date of the contract 合同生效后30天内
20. Besides, you must pay us a running royalty of 3% of the net sales of the licensed products during the period of 10 years after the effectuation of the contract. 另外，合同生效后的10年内，贵方必须支付产品生产提成费，金额为许可证产品净销售额的3%。

29. Buying Know-how
（购买专门技术）

Mr. Brown is a businessman representing a French textile company and Mr. Huang is a manager of China Textile Import & Export Company. The Chinese party wants to buy the French party's know-how, so the two parties are negotiating about the price.

Huang: We'd like to buy your company's know-how.

Brown: Buying the know-how is better than the right to use the patent.

Huang: Why?

Brown: Because the know-how tells all the details of how to manufacture the equipment and buying the know-how will be capable of contributing to advancement of our scientific and technical level.

Huang: Then how much will you charge?

Brown: Four times the price for the patent.

Huang: That's too high.

Brown: Oh, just the opposite, buying the know-how will be much cheaper than making the equipment with our patent.

Huang: I'm afraid that your price is higher than I expected. Is it possible for you to reduce it?

Brown: I think the price is reasonable.

Huang: If in that case, there is hardly any need for further discussion. We might as well call the

deal off.
Brown: Well, for friendship's sake, we can consider reducing the price further by 8 percent.
Huang: A cut of 10 percent will be more realistic.
Brown: I think it is unwise for either of us to insist on his own price.
Huang: What's your proposal?
Brown: We meet each other half way in order to narrow the gap.
Huang: Right, you have persuaded me to agree to your terms.

Notes

1. buying know-how 购买专门技术
2. be capable of contributing to advancement of our scientific and technical level 能够为我们的科技水平的提高作出贡献
3. four times the price for the patent 是专利价格的四倍
4. just the opposite 恰好相反
5. We might as well call the deal off. 我们也可能取消这笔交易。
6. for friendship's sake 从友谊的角度考虑
7. We meet each other half way in order to narrow the gap. 为了缩小我们之间的鸿沟，我们各让一半。

30. Transfer the Right to Use the Patent
（转让专利使用权）

Mr. Smith is the business representative of American Electric Appliance Company and Mr. Wang is the representative of China Hua Mei Electronic Company. The Chinese Party needs the American party to supply equipment, but the latter thinks the manufacturing cost is too high and would like to transfer the patent. The two parties are negotiating about patent license, royalty, initial down payment, etc.
Smith: It's the third time of negotiations.
Wang: Yes, we'd like you to provide us with the equipment.
Smith: I'm afraid that we can't because the production costs have risen a great deal and we are now losing money. However, we're quite willing to transfer the patent.
Wang: Well. It seems that we'll only consider buying the patent. In what form will you transfer the patent?

Part Two
Technology Transfer 技术转让

Smith: We'd like to transfer the right to use the patent in the form of license.

Wang: But the license only gives one the right to manufacture the equipment. What about the technology not included in the patent?

Smith: Don't worry. We'll provide you with all the information and also five technicians needed to manufacture the equipment.

Wang: OK. How long will you allow us to use the patent?

Smith: Five years.

Wang: Well, how much will you ask for?

Smith: We hope you will pay us $ 30,000 as initial payment for buying the production rights from us, and 3% of the sales price on each product sold.

Wang: I'm afraid that we can't accept that. We hope you can reduce to 1% of the sales price on royalty.

Smith: It would not seem proper to do so. What about 2%?

Wang: Well, yes.

Wang: I think you should ensure that the technology provided is complete, correct, effective and capable of accomplishing the technical targets specified.

Smith: Of course. Meanwhile, in accordance with the scope and duration agreed upon by us, you shall undertake the obligation to keep confidential, all the technical secrets contained in the technology provided by us, which have not been made public.

Wang: You needn't worry about it. The technology transferred to us will be kept confidential and not let out or passed on to a third party. We hope you shall train our workers to understand and use the know-how, instructions and other technical data and information for the provided technology.

Smith: Acceptable. I think half a year on-the-job training will be enough for the workers to master the skills. But you shall provide us with adequate facility and the necessary tools to our technicians of rendering technical assistance.

Wang: All right. All the costs for them shall be born by you, is that OK?

Smith: Oh, yes, we agree.

Wang: The duration of the contract shall conform to the time needed by us to assimilate the technology and shall not exceed five years.

Smith: It sounds reasonable. How shall we settle disputes?

Wang: All disputes and differences of any kinds arising from the execution of this contract shall be settled amicably by the parties, or shall be submitted to the Chamber of Commerce, London, for conciliation in order to settle the dispute in an amicable manner under the rules adopted by said Chamber.

Smith: We agree with you.

Wang: Anything else you want to bring up for discussion?

Smith: No, nothing else, Mr. Wang.

Wang: I'll have the contract amended and sent to your hotel for you to look over tomorrow afternoon. What do you think?

Smith: That's very nice.

Notes

1. Transfer the Right to Use the Patent 转让专利使用权
2. What about the technology not included in the patent? 没有包括在专利中的技术怎么办呢?
3. 3% of the sales price on each product sold 每出售一件产品就支付给我们其售价的3%作为提成费
4. It would not seem proper to do so. 这样做似乎不太恰当。
5. I think you should ensure that the technology provided is complete, correct, effective and capable of accomplishing the technical targets specified. 我认为你们应该确保所提供的技术是完整的、无误的、高效率的而且是能达到规定的技术目标的。
6. Meanwhile, in accordance with the scope and duration agreed upon by us, you shall undertake the obligation to keep confidential, all the technical secrets contained in the technology provided by us, which have not been made public. 与此同时，根据我们商定的范围和期限，你们应该承担保密的义务，包括在我们所提供技术中的所有技术秘密，因为这种技术一直没有公开。
7. The technology transferred to us will be kept confidential and not let out or passed on to a third party. 转让给我们的技术将被严格保密，绝不会泄露或转让给第三方。
8. to understand and use the know-how, instructions and other technical data and information for the provided technology（使他们）知晓和使用所提供技术的专门知识、说明书和其他技术资料与信息
9. to master the skills 掌握这些技能
10. But you shall provide us with adequate facility and the necessary tools to our technicians of rendering technical assistance. 但你方应向我方的技术人员提供足够设施和进行技术援助所需的工具。
11. The duration of the contract shall conform to the time needed by us to assimilate the technology. 合同期限应与我们吸收消化所提供技术需要的时间相符合。
12. How shall we settle disputes? 争执怎样解决?
13. All disputes and differences of any kinds arising from the execution of this contract shall be settled amicably by the parties, or shall be submitted to the Chamber of Commerce, London, for conciliation in order to settle the dispute in an amicable manner under the rules adopted by said Chamber. 在执行本合同期间产生的所有争执和分歧应由双方友好解决。否则，应提交伦敦商会调解，以便按该商会通过的规则以友好方式加以解决。
14. Anything else you want to bring up for discussion? 你还想提出另外的问题进行讨论吗?
15. I'll have the contract amended. 我将对合同作些修改。

Part Three 3

International Engingeering Project Cooperation

国际工程项目合作

31. International Engineering Project Contract and Service Cooperation（1）
（国际工程项目合同及服务合作之一）

　　Prof. Gao（G）and a student Huang Lan（S）are discussing engineering project contract and service cooperation.

S：Prof. G, would you please give me a general introduction to the engineering project contract and service cooperation?

G：No problem. The international engineering project contract is a comprehensive international cooperation in economy and technology, including machines and equipment, technology, capital and service, under which one party (the contractor) shall undertake the responsibility for fulfilling the building of a certain project item, and the other party (the employer) shall offer necessary working conditions and give the acceptance to the project, and pay the agreed amount of the price value and the service reward.

S：What are the key points of engineering project survey and evaluation?

G：Generally speaking, there are 6 aspects. ① First of all, an all-round hydrological and geological survey must be made. Then on the basis of the investigation, a technology appraisal and an economic evaluation of the project item proposed to be built must be made for further feasibility study. ② Designs of engineering project. Sometimes they are referred to as project designs, which mainly include the conceptual design, basic design and detailed design, corresponding to the initial design, technology design and specific design for construction that are widely adopted in practice in some countries. ③ Supply of technology. Not only common technology, but also patent technology and know-how are included. The former is usually provided in technology consultation and personnel training, the latter may be patent technology or know-how that is needed in engineering project.

S：Then the other three?

G：Well. ④ Supply of machines and equipment. They may be wholly or partly supplied by the contractor. Sometimes, the contractor is also responsible for supplying raw materials in terms of the stipulations in the contract signed by both sides. ⑤ Construction and installation. To fulfill these tasks, much work must be done in many aspects, including the dispatch of experts and workers, the supply of machinery, and their specific use and operation. ⑥ Initial operation. After the whole engineering project construction has been completed, the contractor is responsible for conducting a test run of the production equipment to see whether the trial run or sample production comes up to the standard, or meets the requirements prescribed in the contract. The contractor may contract for the whole engineering project construction, or may contract for a part or several parts of the construction under the monopoly contract of split-phase engineering. If the contractor contracts for the construction of the project completely, he may again contract several parts out to other subcontractors.

S: According to the modes of business operation, how many types of the international engineering project contracts may be classified into?

G: They may be classified into: ① general monopoly contract; ② monopoly contract of split-phase engineering; ③ split-pals engineering contract; ④ transferring engineering contract; ⑤ joint management contract; ⑥ joint investment contract; ⑦ contract of split-phase service of engineering.

S: According to the contents of engineering project, how many types may they fall into?

G: They may fall into: ① engineering project construction contract; ② installation of equipment contract; ③ engineering project drawing contract; ④ semi-turn-key engineering contract; ⑤ turn-key engineering contract; ⑥ product-in-hand engineering contract.

S: According to the modes of valuation, how many types of them may be classified into?

G: There are three: ① total price contract; ② unit price contract; ③ cost and commission contract. All the businesses mentioned above are usually concluded through bidding and negotiations.

S: What are the basic characteristics of international engineering project contracts?

G: I think there are six: ① complicated process; ② high-level technology and management; ③ long time for construction; ④ great amount of investment; ⑤ big risks; ⑥ generous profits.

S: Prof. G, would you please give us some more knowledge about the contract or agreement?

G: With pleasure. Generally speaking, there are 3 parties to the contract or agreement. It is the contractor's main duty to fulfill the construction task of the engineering project in light of the requirements of quality, quantity and time limitation at his own expenses and risks, and his main right is to get the price value and reward. The client or employer is mainly responsible for providing necessary working conditions and paying the price value and reward for the project construction in line with the agreed terms in the contract, and he has the right to obtain the ownership of the engineering project.

S: In activities of international economic and technological cooperation, is a supervisory engineer normally appointed by the employer, of which the contractor must be advised by a written document?

G: Surely. The supervisory engineer's main right and duty are to draw up implementation plans for the engineering project construction, issue orders of going into operation, make engineering project draw for and give technical instructions to the construction, prepare detailed specifications list, engineering project operation capacity or work quantity form and other contractual documents, supervise the implementation of the whole project, make acceptance of the completed project, issue the certificate of completion or the certificate of payment. In addition to those mentioned above, sometimes, the supervisory engineer may act as a just intermediary between the employer and the contractor if any small dispute occurs.

高新技术、技术转让与国际工程合作

Notes

1. give the acceptance to the project 接收工程
2. the service reward 服务报酬
3. conceptual design 概念设计
4. corresponding to the initial design 相应的原创设计
5. the dispatch of experts and workers 派遣专家和工人
6. under the monopoly contract of split-phase engineering 根据总承包合同的分项工程（承包）
7. general monopoly contract 工程总承包
8. split-pals engineering contract 部分工程承包
9. turn-key engineering contract 总承包合同
10. generous profits 可观的利润、丰厚的利润
11. draw up implementation plans 起草实施计划
12. make engineering project draw for and give technical instructions to the construction 准备工程建设图纸并给予技术指导
13. prepare detailed specifications list, engineering project capacity or work quantity form 编制规格明细表和工程量表

32. International Engineering Project Contract and Service Cooperation (2)
（国际工程项目合同及服务合作之二）

The conversation between Prof. Gao and the student is going on.

S: It is said that big or complicated technical items, financed by international banking facilities or assisted by the United Nations and international organizations are often contracted out by invitation for bids. Is it really so?

G: Yeah. To call for bids, the promoter will have to organize a special committee on invitation of bidding to be in charge of the invitation to bid. At the same time, a supervisory organ is also set up to supervise the work.

S: What are the principal contents of bid documents?

G: Bid documents, namely inquiry documents, should be carefully prepared for invitation to bid. They are regarded as the foundations for calling for bids and mainly include: ① invitation for tender; ② Information for or instructions to tenderers; ③ formation of application for tenders and attachments; ④ terms of contract; ⑤ formation of contract; ⑥ general description; ⑦ form of general engineering operation capacity or general bill of work quantities; ⑧ drawings and attachments; ⑨ basic data; ⑩ additional data list; ⑪ Summary of bid documents.

S: What are the terms of contract?

G: Well. Terms of contract constitute the chief contents in the contract document of international engineering project, which prescribe the legal relations among the employer, the contractor and the supervisory engineer, and their rights and duties respectively. The following fundamental clauses are usually included in the terms of contract: ① rights and duties of supervisory engineer and his representatives; ② clause of assignment or subcontract; ③ general duties of the contractor; ④ clause of special natural conditions and artificially imposed obstacles; ⑤ clause of the contractor's supervision; ⑥ clause of engineering protection; ⑦ clause of insurance; ⑧ clause of tests; ⑨ clause of shutdown; ⑩ clause of delay in completion; ⑪ penalty for late completion; ⑫ loss of working time; ⑬ clause of maintenance; ⑭ working charges and modifications of engineering project; ⑮ claim for supplementary fees by the contractor; ⑯ clause of payment; ⑰ breach of contract by the contractor; ⑱ breach of contract by the client; ⑲ arbitration; ⑳ clause of special risks.

S: What about the base price limit on bids?

G: This is a good question. It is an important component of the procedure of invitation to bid, and a preliminary contract price involves very complicated accounting, worked out by the employer. To determine the base price limit on the basis of practical and reasonable valuation is of great significance for the benefit to and success on the invitation of bidding. The base price limit on bids constitutes a top secret part of all the prepared data and not in any case can it be let out to any other person or unit, otherwise the whole work will be meant a failure.

S: Then the technical specifications?

G: Technical specifications are technical targets laid down by the promoter in international engineering project contract, which must be complied with by the contractor in construction, and in the supply of equipment and materials to be needed. These detailed technical requirements and the construction drawings together form the complete technical documentation for engineering project.

S: How many kinds of invitation to bids are there in international engineering project?

G: There are a lot of types available for invitation to bids. The most popular adopted types are shown below: ① as per territories: They may be classified into international tender offering and domestic tender offering; ② as per degrees of competitiveness: They may be classified into competitive tender offering, noncompetitive tender offering or negotiating tender offering, unlimitedly-competitive tender offering, limitedly-competitive tender offering or selective tender offering or inviting for tender, exclusive tender offering, two-stage tender offering or tender offering consisting of open and selective stages; ③ as per ranges of engineering project: they may be classified into tender offering for turn-key engineering project, tender offering for product-in-hand engineering project, tender offering for construction of engineering project and installation of equipment, parallel tender offering, sequential tender offering, tender offering for purchase of equipment and materials; ④ as per modes of valuation of engineering project: they may be classified into tender offering for fixed-total-price engineering project, tender offering for cost-and-commission engineering project.

Notes

1. international banking facilities 国际银行信贷业务机构（国际银行业服务或贷款项目）
2. invitation for bids 招标
3. to call for bids 要招标（邀标）
4. the promoter 发标人，招标人
5. a supervisory organ 监督机构
6. invitation for tender 招标书
7. information for or instructions to tenderers 投/招标人须知
8. formation of application for tenders and attachments 投/招标书格式与附件
9. form of general engineering operation capacity or general bill of work quantities 工程量总表
10. drawings and attachment 工程设计图纸和附件
11. artificially imposed obstacles 人为障碍
12. clause of shutdown 停工条款
13. clause of delay in completion 延期完成工程条款
14. penalty for late completion 推迟完工处罚条款
15. loss of working time 误工条款
16. working charges and modifications of engineering project 工程运转费用与修改
17. claim for supplementary fees by the contractor 承包商索取追加费用条款
18. the base price limit on bids 标底
19. which must be complied with by the contractor in construction 在施工中，工程承包商必须遵从
20. as per territories 根据地域划分
21. international tender offering and domestic tender offering 国际招标和国内招标
22. unlimitedly-competitive tender offering 无限竞争性招标
23. selective tender offering or inviting for tender 选择性招标或邀请性招标
24. exclusive tender offering 排他性招标
25. tender offering for turn-key engineering project 总承包工程招标
26. parallel tender offering, sequential tender offering, tender offering for purchase of equipment and materials 平行招标、序列招标、招标采购设备和材料

Part Three
International Engineering Project Cooperation 国际工程项目合作

33. International Engineering Project Contract and Service Cooperation (3)
(国际工程项目合同及服务合作之三)

S: If the tender offering is open to the public, should the organs in charge of invitation of bidding make use of every possible means to publish the notice inviting for bidding so as to attract contractors to bid?

G: Certainly. After the contractors have got the information, they must make application for bidding if they are interested in the engineering project contract. But only those who have passed through prequalification may buy bid documents, prepare tender documents and compete with other contractors for winning the bid.

S: What should tender documents usually consist of?

G: Generally speaking, they include the following contents: ① form of tender and attachments; ② tender guarantee; ③ price and expenses list; ④ job schedule of engineering project; ⑤ time and method of payment; ⑥ construction scheme; ⑦ construction organization and biographical records of major managerial personnel to be appointed; ⑧ labor arrangement program; ⑨ all relative data, such as agreement, terms of contract, drawings of engineering project, description and other lists and materials. Tender documents usually contain all the instruments for tender mentioned in the bid documents except the summary of bid documents, but having already been filled by the bidders, an amending bid document, a question sheet and written answer, a letter of guarantee for tender or tender bond, a letter on submission of tender.

S: What is the most important work for bidders to do in the process of preparation for bidding?

G: Well, according to the items listed in bid documents, bidders must fully calculate, work out and fill in quotation data including the unit price, line-item price, total price and other items needed to be filled out. Since the main part of working out bid documents is to calculate the bid price and fill in the documents with it, it is usually referred to as bid quotation or bidding.

S: How can the bidders work out tender documents well?

G: They must first of all make themselves acquainted with the bid documents, especially the general description of the engineering project, technical specification, construction drawings, general engineering operation capacity, machines, equipment and raw materials to be needed, labor service, and so on. If they may provide better schemes of design and construction for the engineering project, they are required to send to the employer a new detailed drawing and description when the quotation is made. Two major types of tender are usually adopted in international engineering project contract and service cooperation. If the bidder independently enters a bid, it is called exclusive tender; if two or more than two contractors organize a new business entity to submit tender together, then it is called joint tender.

S: After submission of tenders has been ended, what should the employer usually do?

G: Well, the employer should hold a ceremony for bid opening within the stipulated time limit at the named place. The employer usually entrusts the institution on invitation of bidding or an advisory company with the responsibility for bid opening within 90 days after the closing date of tender. In terms of the specific conditions three modes may be adopted: ① bid opening in public; ② bid opening limitedly; ③ bid opening secretly.

S: Is evaluation of tender one of the most important and complicated stages in international bidding?

G: All right. After bid opening, the employer, the advisory engineer and the committee on invitation of bidding must make analysis, comparison and evaluation of all the tender documents thoroughly and secretly. It will take most part of the time from the conclusion of bid opening to decision.

S: What are the main contents of the evaluation of tender?

G: There are usually 5 points. ① Comparison of tender price. The work must be done by the employer, the advisory engineer and the committee in evaluation activities. On the basis of all-round comparison of all the basic tender prices and extra tender prices submitted by the bidders, four or five bidders are usually selected, who have made the lowest tender quotations of all the bidders, as the preliminary candidates for the tender discussions. ② Evaluation of contracts and administrations. This is an initial examination carried out before the thorough evaluation is conducted, so as to pick out the tender documents which are in keeping with the requirements set in the bid documents for bidding.

S: Then the other three points?

G: Well. ③ Technological evaluation of tenders. After the comparison of tender prices and evaluation of contracts and administrations, the employer must entrust several experts with the technological evaluation of the tender documents which have been picked out in the preliminary examination. The further technological evaluation mainly centers on the general description, technical drawings, general designs and schemes of construction, control of engineering quality, construction administration, technical ability and service, and etc. ④ Commercial evaluation of tenders. The commercial evaluation of tender mainly includes the contents in finance, accounting, cost control, and economic management, and so on. ⑤ Analysis of engineering risks. Before making the choice of the successful bidder, the employer must also take into account all the risks which may possibly occur in the engineering construction.

S: Are tender discussions and tender decision carried out between the employer and the bidders chosen after the evaluation of tenders, including the technical reply and price negotiations?

G: Surely. Tender decision shall normally be made within the period of three to six months. Then the promoter should send a notice of award to the winning bidder. The last work for both parties to do is to conclude an agreement of contract to determine the terms and conditions of contract and the rights and duties of the promoter and the contractor. At the same time, the contractor shall establish with his banker a letter of performance guarantee and submit it to the employer as an economic guarantee.

S: What are the main aspects and contents of technical service and common labor service in international service cooperation?

G: The types and fields of service cooperation are continuously expanding with the social development and the changes of international economic relations. But on the whole, they can be divided into two types, technical service and common labor service. The cooperation in this respect is usually carried out by signing a contract of service cooperation by the parties concerned. The main points are as follows: ① service cooperation of international engineering project contract; ② simple service cooperation; ③ enterprise migration.

S: What contents should be clearly stipulated in the contract of service cooperation?

G: There are altogether seventeen: ① dispatch of personnel; ② duties of both parties; ③ formation of expenses, including wages, overtime pay, allowances, traveling expenses, office expenses; ④ working days and holidays; ⑤ working conditions; ⑥ break of work; ⑦ labor protection; ⑧ mode of payment; ⑨ replacement and dismissal of personnel; ⑩ performance guarantee; ⑪ confidential; ⑫ equipment and tools; ⑬ medical conditions and social insurance; ⑭ assignment; ⑮ disputes and arbitration; ⑯ force majeure; ⑰ governing laws and regulations.

Notes

1. passed through prequalification 通过资格审查
2. form of tender and attachments 投标书和附件
3. job schedule of engineering project 工程进度表
4. biographical records of major managerial personnel to be appointed 委派的主要管理人员的个人简历
5. all the instruments for tender 投标的所有文件（契约）
6. tender bond 投标保证金
7. a letter on submission of tender 投标致函
8. bidders 投标人、投标商
9. for bidding 投标
10. line-item price 分项价格
11. bid quotation 报标或开价
12. a notice of award to the winning bidder 授予中标人通知书
13. The contractor shall establish with his banker a letter of performance guarantee. 承包商必须向其开户行提供工程实施的履约保证书。
14. simple service cooperation 单纯劳务合作

15. enterprise migration 企业移民
16. overtime pay 加班工资
17. allowances 津贴、补助费
18. assignment 转让（指财产、权利的转让）
19. force majeure 不可抗力
20. governing laws and regulations 适用法律法规

34. A Turn-key Project
（总承包工程项目）

（1）Construction Work

Mr. Smith, the contractor, is discussing with Mr. Pan the proceeding of the construction work.

S: I'm very pleased that my firm has been awarded this contract. To ensure timely completion of the project, I suggest that we use local building equipments and materials.

P: We agree to make full use of local resources.

S: Good. And we intend to use local labour too. Still, we'd like to reserve the right to engage foreign firms as well.

P: Of course you may, if the engagement of foreign subcontractors conforms to the current laws and regulations.

S: We'll see to that. About the completion time of the construction, we will be able to finalize it when the delivery time of local building materials and equipments is fixed.

P: That can be fixed in one or two days.

S: Thanks. And the construction period also depends on the time when you hand over the construction site to us and start to finance the construction work.

P: The handing over of construction site will be done on time, and financing will start as scheduled.

S: I'm glad to hear that. Now, about the third firm Engineer, have you appointed it yet?

P: Yes, we've signed the contract with ABC Ltd, London, which will act as the Engineer on construction site.

S: You've chosen a competent Engineer to represent the Customer on the site to decide all technical matters.

P: Yes, the Engineer will act in our interest within the limits of the Contract. He will have access to all work done on the site, especially to the work which cannot be checked after completion. Any work that is proved to be done improperly will have to be re-done at the Contractor's expense.

S: I assure you, we'll observe the contract stipulations to the letter.

(2) Subcontracting

S: I've studied your request to shorten the period of construction and have work on the site finished before the freezing season. I think this can be done only if, first, you undertake to provide local building materials and equipment in time, and second, to recommend the best local civil engineering firm as subcontractor to help in the work.

P: We'll surely provide materials and equipment required by you on schedule. As to a subcontractor I recommend the No. 1 Construction Co. here. I'll arrange for you to meet each other tomorrow.

S: Thank you, Mr. P. But we'd like to reserve the right to engage a foreign firm as our subcontractor.

(Two weeks later.)

S: Mr. P, it's high time to solve the problem of delay in civil engineering work. Otherwise, we won't be able to complete the construction on time.

P: I can see we are behind schedule with the work.

S: I'm afraid the firm recommended by you cannot cope with the problems.

P: Are you sure? Maybe they are having temporary difficulties with a new job.

S: I'm sorry it's not so. It's obvious that the firm is incapable of meeting the contract date.

P: If so, we agree to engage another subcontractor.

S: I think EB Corp. is the best choice.

P: Agreed. I only hope that firm will not let you down.

S: I don't think they will. And, what about the extra expenses involved when signing a new contract, now that the rates for civil work have gone up lately?

P: The contract price provides for unforeseen expenses like this.

S: That's true. But the additional expenses may well exceed the amount provided in the contract. I hope you will be present at the talks with the new subcontractor. The representative of that firm will be here in two days.

P: I'll report all this to the president and let you know as soon as there is any news.

Notes

1. my firm has been awarded this contract 我方公司能够承包（获得）这项合同工程
2. to ensure timely completion of the project 为了保证按时完工
3. We agree to make full use of local resources. 我同意充分利用当地资源。
4. intend to 有意向（准备）
5. Still, we'd like to reserve the right to engage foreign firms as well. 但我们仍要保留聘请外国公司参加施工的权利。
6. if the engagement of foreign subcontractors conforms to the current laws and regulations 如雇聘外国分包商参加施工，需符合（我国）现行法律和法规

7. We'll see to that. 我们会注意做到。
8. About the completion time of the construction, we will be able to finalize it when the delivery time of local building materials and equipment is fixed. 关于完工的时间，只有当提交本地建筑材料和设备的时间确定后，我们才能最终定下来。
9. the construction period also depends on the time when you hand over the construction site to us and start to finance the construction work 施工期还取决于贵方将施工现场移交给我们以及开始为施工提供资金的时间
10. as scheduled 按计划
11. Now, about the third firm Engineer, have you appointed it yet? 现在谈谈第三家公司工程师的问题，你们是否已经委派？
12. We've signed the contract with ABC Ltd, London, which will act as the Engineer on construction site. 我们已与伦敦 ABC 有限公司签订了合同。该公司将担任工地工程师。
13. You've chosen a competent Engineer to represent the Customer on the site to decide all technical matters. 你们已选择了一位合格的工程师，在工地上代表客户（订货方）决定所有技术问题。
14. will act in our interest within the limits of the Contract 将在合同范围内，代表我方利益行事
15. He will have access to all work done on the site, especially to the work which cannot be checked after completion. Any work that is proved to be done improperly will have to be re-done at the Contractor's expense. 他将能进入工地所有完工的工作，特别是完工后不能检查的工作。任何经证明没有完成好的工程项目必须返工，费用由承包商负担。
16. I assure you, we'll observe the contract stipulations to the letter. 我向你保证，我们一定严格按合同规定书施工。
17. have work on the site finished before the freezing season 在冰冻期以前完成工地工程
18. to recommend the best local civil engineering firm as subcontractor to help in the work 推荐当地最好的土建工程公司作为分包商，协助施工
19. It's high time to solve the problem of delay in civil engineering work. 是解决土建工程延误问题的时候了。
20. behind schedule with the work 落后工程进度时间表了
21. the firm recommended by you cannot cope with the problems 贵方推荐的公司处理不了工地上的问题
22. incapable of meeting the contract date 不能按合同规定日期完工
23. that firm will not let you down 那家公司不会让你失望
24. Now that the rated for civil work have gone up lately? 近来的土建工程费用可能已上涨了？
25. The contract price provides for unforeseen expenses like this. 类似的未能预见（意外）的费用，已包括在合同金额内了。
26. But the additional expenses may well exceed the amount provided in the contract. 可这笔额外的开支可能会大大超过合同所包括的金额。

Part Four 4
Inviting Bids and Bidding
招标投标

35. Inviting Bids and Bidding
（招标投标）

A: B, our government is going to invite tenders for the Changjiang Three Gorges Project. Would you like to take part in?

B: Yes, participation in tenders is in our line of business, but we would like to know what we shall have to do if we agree to send our bid.

A: As it stands, in addition to the bid you are to submit information on cost, construction time and the volume of works concerning the projects already constructed by you.

B: We'll try to do it without delay, but we would like to know the requirements of the tender committee.

A: Certainly, we shall get a complete set of tender documents for you and you will be able to study the requirements. The expenses involved will be charged to your account, though.

B: Well, no problem. By the way, must we guarantee in any way of our participation in the tender?

A: You will have to pay "earnest money" to guarantee your participation till the end of the tender.

B: That's quite fair. What are our chances of success?

A: We know that you have rich experience in this field and that you render technical assistance on favorable terms. I think you may win the tender.

B: I hope so. We must reconsider your offer and we shall give our reply in the near future.

A: Well, we look forward to your reply as soon as possible.

B: Thank you for your information.

A: You're welcome.

Notes

1. invite tenders 招标
2. the Changjiang Three Gorges Project 长江三峡工程
3. participation in 参加
4. in our line of business 是我们的业务专长；我们有业务优势
5. As it stands, in addition to the bid you are to submit information on cost, construction time and the volume of works concerning the projects already constructed by you. 是这样的，除了寄标书外，你方应把曾承建过的有关工程项目的费用、时间、数量等信息一并寄来。
6. a complete set of tender documents 一套完整的投/招标文件
7. The expenses involved will be charged to your account, though. 但由你方承担相关费用。

8. Must we guarantee in any way of our participation in the tender? 我方必须以某种方式担保参加投标吗?
9. earnest money 保证金
10. render technical assistance on favorable terms 以优惠的条件提供技术援助

36. Different Opinions about Bid Invitation
（关于招标的不同意见）

Andros—A, Ferguson—F, Azevedo—Z, Silva—S

A: The National Development Agency of Tanaku is going to start a major new project which has the full support of our government and the President himself. It will bring prosperity to the whole country, but especially to the southern region. The southern region of Tanaku is dry, with little cultivation. However, a short time ago, we discovered that the sand of one beach in the south is almost pure quartz, suitable for processing into silicon. We have decided not to sell the sand, but to develop our own silicon processing plant near the beach, with a factory to manufacture solar energy panels using silicon wafer.

S: We expect that several international companies will submit tenders. But we will require them to cooperate with local construction firms, here in Tanaku. We will use local firms as much as possible.

F: But can a country like Tanaku control the project when the finance has to be raised internationally?

Z: Of course, we need finance, training and know-how from abroad. So we are going to ask our friends from overseas to be our partners. But we will repay our debts in full from the profits we make. Then the people of Tanaku will own the factories and plants themselves.

A: Dr. Azevedo was optimistic. But other people at the conference saw difficulties ahead. As I understand from your outline, Tanaku will need only one third of its silicon for the manufacture of solar panels.

Z: That is correct. However, we hope to extend the factory in the next ten years or so.

F: But during that time, can you be sure of a market?

Z: According to our consultants, manufacturers in other countries will need large quantities of high-grade silicon. In view of this world-wide demand, we have no worries about the market.

F: It's a very big project, Andros. Some people think it's too big for the National Development Agency to handle. There are too many risks. What do you say? If Tanaku produces its own silicon, it can make solar panel. But it will also have to sell twelve thousand tons of silicon every year for several years. How can you be sure of a market? Are other silicon manufacturers going to stand aside and allow Tanaku into the market? Surely they will increase their own output to meet the new demand? Don't you agree?

A: I will tell you my opinion about the silicon project, Mr. Ferguson. For some time now, our President has been influenced by politicians from the south. This project will be too big for our country. We are biting off more than we can chew.

F: More than you can chew? Yes.

A: Yes. How do we know that we can sell all that silicon? Tell me that!

F: You have a very good point. So why not sell the quartz sand to Pansil?

A: You said why not to sell the quartz to Pansil?

F: Yes, yes! That's excellent! Let Pansil buy the quartz and take the risks.

A: And let them take all the risks!

F: What a wonderful idea! Will you speak to the President about it?

A: Yes. I'll speak to him tomorrow. It's a question of making the right choice.

Notes

1. ... is going to start a major new project which has the full support of our government and the President himself. ……将启动一个大的新工程，它会得到我们政府和总统本人的全力支持。

2. bring prosperity 带来繁荣

3. with little cultivation 很少耕种农作物

4. ... with a factory to manufacture solar energy panels using silicon wafer. ……再建一家工厂，用硅晶片制造太阳能电池板。

5. submit tenders 参加投标

6. But can a country like Tanaku control the project when the finance has to be raised internationally? 但是当资金必须在国际上筹集时，像塔拉库这样的国家能控制这种项目吗？

7. Of course, we need finance, training and know-how from abroad. So we are going to ask our friends from overseas to be our partners. 当然，我们需要国外的资助、培训和技术支持，所以我们将请我们的海外朋友作为我们的合作伙伴。

8. ... repay our debts in full from the profits we make. ……我们将用赚取的利润偿还全部债务。

9. optimistic 乐观的

10. ... saw difficulties ahead. ……看到前面的困难还有不少。

11. As I understand from your outline ... 就我对你们草案的理解，……

12. ... we hope to extend the factory in the next ten years or so. ……我们希望经过10来年的时间扩建工厂。

13. high-grade silicon 高品质硅

14. Are other silicon manufacturers going to stand aside and allow Tanaku into the market? 其他硅制造商会袖手旁观，让塔拉库进入市场吗？

15. For some time now, our President has been influenced by politicians from the south. 在目前的

一段时间里，我们的总统受到南方政治家的影响。
16. We are biting off more than we can chew. 我们力不从心。（我们吃得太多而咀嚼不了。）

37. To Find a Bidder Who Can Guarantee a Market
（寻求有市场保证的投标商）

Azevedo—Z, Silva—S, Christine—C, James Clarke—J

C: Do you think that many firms are going to bid for this project?
J: I don't know. Frankly, I think Pansil has an advantage over the rest.
C: Oh? Why?
J: Well, because it's a very big group and can make a turn-key bid for the whole project. And that's what I think they're going to do.
C: But Mr. Silva has said he will also consider package bids for various parts of the project.
J: But a turn-key bid is more attractive for the customer. Then you deal with one company, and that company deals with the problems of its partners.
(Now Silva and Azevedo come and explain the reason for their sudden visit.)
Z: I've tried to reassure the President. I told him there's no cause for alarm. But he's very worried. He says we must find an answer to Andros.
S: He's a politician. He has a big following in the north.
Z: The President can't ignore Andros. The man is too powerful.
J: Why does he want to stop the project?
Z: Because it will develop the south, not the north. So he says that Tanaku will not be able to sell its extra silicon.
S: I wonder where he got that idea.
J: I'm sure it's not true. The other silicon manufacturers can't keep Tanaku out of the market.
Z: But that's just what Pansil will do, according to Andros.
C: Why should they do that?
Z: Because they want to buy the sand and convert it into silicon themselves. They don't want us to have our own silicon plant.
C: Can they keep Tanaku out of the market?
J: Well, it's possible.
Z: Anyway, Andros will say they can, on every possible occasion. He will cause trouble. We are depending on international banks for aid, and bankers are cautious people. If they think our government is divided they will refuse to help.
C: But then what will happen?

S: We'll have to forget about our own project. We'll just have to sell the quartz sand to Pansil.

C: Which is what they've always wanted.

Z: No! We must go ahead with our project.

S: But what can we do?

C: It's quite simple. You must find a bidder who can guarantee a market for the extra silicon. That's the answer to Andros.

S: Does such a company exist?

J: Well, no-apart from Pansil. They're big enough to provide their own market. They have companies in the group which manufacture many silicon-based products.

S: Well, it seems to me that Pansil with the help of Andros, are forcing us to do exactly what they want.

C: We must find a bidder who will guarantee a market for the extra silicon.

J: I've already told you, there aren't any other groups that big.

C: That's right, there aren't. But perhaps there will be.

S: What do you mean, Christine?

Notes

1. Do you think that many firms are going to bid for this project? 你认为将有许多公司为这个项目投标吗？

2. ...can make a turn-key bid for the whole project. ……能为这个项目提出总承包工程投标。

3. ...he will also consider package bids for various parts of the project. ……他也会考虑对工程的各个部分（子项目）进行组合（一揽子）投标。

4. ...that company deals with the problems of its partners. ……那家公司可帮助其合伙人解决难题。

5. I've tried to reassure the President. 我已尽力使总统消除疑虑。

6. I told him there's no cause for alarm. 我告诉过他没有理由惊慌。

7. He has a big following in the north. 他在北方有一大群追随者。

8. The President can't ignore Andros. 总统不能忽视安德罗斯的意见。

9. The man is too powerful. 这人很有影响力。

10. ...can't keep Tanaku out of the market. ……不会把塔拉库排挤在市场大门之外。

11. But that's just what Pansil will do, according to Andros. 但根据安德罗斯的说法，潘西尔公司就是要这样做。

12. on every possible occasion 非常可能

13. He will cause trouble. 他将会制造麻烦。

14. ...and bankers are cautious people. ……银行家都是谨慎小心的人。

15. If they think our government is divided they will refuse to help. 如果他们认为我们政府有分歧意见，就会拒绝帮我们。

16. Which is what they've always wanted. 这正是他们朝思暮想的美事。
17. We must go ahead with our project. 我们必须坚持发展我们的工程。
18. You must find a bidder who can guarantee a market for the extra silicon. 你必须找到一个投标商来担保多余硅的销售市场。
19. Well, no-apart from Pansil. 嗯，不存在，我们离不开潘西尔公司。
20. ... manufacture many silicon-based products. ……生产许多硅制品。
21. Well, it seems to me that Pansil with the help of Andros, are forcing us to do exactly what they want. 哦，照我看来，潘西尔公司借助安德罗斯向我们施加压力，想让我们完全按照他们的想法办事。

38. It is Fairy Normal at This Stage for Bidder to Offer Extras
（在投标的这个阶段投标商提供更优惠的条件很正常）

Watanabe—W, Silva—S, Azevedo—Z, Ferguson—F

W: The Kalamaro Plateau, I understand, is a problem area for Tanaku. Now, your new power station for the silicon project will be sited here, on the west coast. A transmission line will run south down here, in order to service the silicon plant, the solar panel factory and the new town at Yatu Beach. Gentlemen, in order to demonstrate our good will towards Tanaku, we propose to offer your country an additional benefit. We will build another transmission line from here, which will provide a source of electrical power on the Kalamaro Plateau.

S: It would provide power to the plateau? At no extra cost to us?

W: Quite so. At no extra cost to you.

Z: Of course, I agree that we should be grateful for the offer. But the value of an extra transmission line depends on what we can do with the power when we've got it.

F: Forgive me, Dr. Azevedo, but surely there are a number of industries that can be set up?

S: A cement works, for instance.

F: A good idea, Mr. Silva. That would provide employment for people in the area.

S: We've already looked into the possibility of a cement works. A preliminary survey has been done, and electrical power on the plateau would make that project possible. Some sort of development is needed urgently, in order to provide the people with employment.

Z: I agree in principle. But I suggest we ask our consultant to evaluate this very interesting offer from Pansil. Let him investigate the cement works project and advise us on the best course of action. Meanwhile, thank you very much, Mr. Watanabe.

(Ferguson and Watanabe leave the NDA together.)

F: I think that went well. Silva is very keen on the cement works. It's rather a good idea of mine-

power for the plateau.

W: We shall see, Mr. Ferguson. You are certainly well informed. It was you who told us about Yatu Beach in the first place. My company is always ready to reward people as they deserve.

F: Thank you.

W: Although we haven't got the contract yet, of course.

F: Neither has anyone else.

W: Quite so.

(Sliva watches Ferguson and Watanabe drive away from the balcony outside Azevedo's office. Azevedo joins him.)

S: I don't understand. Why can't we just accept the Pansil tender?

Z: No, Manuel. We must listen to what James Clarke has to say.

S: But...

Z: And another thing. The Faulkner consortium may come up with an even more attractive offer.

S: But they don't know about the Pansil offer, do they?

Z: No. But it's fairly normal at this stage for bidders to offer extras. And they'll probably learn that Mr. Watanabe has been here. In fact, it would do no harm if they heard that he had been here, Manuel. Then, if Faulkner offered us something extra...

S: We would be pleased to consider it.

Notes

1. The Kalamaro Plateau 卡拉马罗高原

2. ...is a problem area for Tanaku. ……是塔拉库国家一个困难地区。

3. A transmission line will run south down here, in order to service the silicon plant, the solar panel factory and the new town at Yatu Beach. 一条输电线路会向南架到这里，服务于硅工厂、太阳能电池板制造厂和雅图海滩的新城。

4. good will 善意

5. an additional benefit 额外利益

6. At no extra cost to us? 我们不用额外负担费用？

7. But the value of an extra transmission line depends on what we can do with the power when we've got it. 但是，架另一条输电线路的价值取决于我们如何使用这些电能了。

8. We've already looked into the possibility of a cement works. 我们已经调查了建造水泥厂的可行性。

9. A preliminary survey has been done, and electrical power on the plateau would make that project possible. 我们已经做了初步调查，在卡拉马罗高原上建立电站使这项工程的开展成为可能。

10. Some sort of development is needed urgently. 迫切需要某种类型的开发项目。

11. ...advise us on the best course of action. ……就最佳的行动路线（方案）向我们提供咨

询意见。

12. NDA: National Development Agency 国家发展署
13. I think that went well. 我认为事情进展顺利。
14. You are certainly well informed. 你一定信息灵通。
15. My company is always ready to reward people as they deserve. 我们公司一贯做好了准备，要报偿（答）那些值得报答的人们。
16. Neither has anyone else. 别人也没得到合同。
17. Sliva watches Ferguson and Watanabe drive away from the balcony outside Azevedo's office. Azevedo joins him. 阿泽维多走进来搭话，思尔西从阿泽维多办公室阳台上看着费谷森和瓦塔那布开车离去。
18. tender 投标
19. The Faulkner consortium may come up with an even more attractive offer. 福尔克纳财团可能会提出更有吸引力的报盘。
20. ...if Faulkner offered us something extra... ……如果福尔克纳公司给我们提供一些额外的帮助……

39. Your Bid Has Been Accepted
（你方中标了）

Engel—E, Silva—S, James Clarke—J, Azevedo—Z, Davidson—D

E: The solar panels can be set up at any convenient position, near to a well on the Kalamaro Plateau, for instance. The panels are at an optimum angle, to make the most of the sunlight throughout the day.

S: Well, we have plenty of sunlight on the plateau. What has to be done to maintain the panels?

E: Nothing at all. You can wipe the dust off the panels to keep them clean.

S: I'm glad the operation is so simple.

J: Well, it's the way things are going. Perhaps the way they ought to go. Small sources of power, easily available.

E: This electrical motor that works the pump is controlled by a micro-computer. So, it switches itself on and off, as power is available.

Z: There's a computer, in here?

E: Yes. A micro-computer. The whole unit is sealed. And the whole operation is automatic. It needs no maintenance whatsoever.

S: Water. Irrigation. That is the answer!

D: Mr. Silva. As part of our bid for the silicon project, we are proposing that Micrel will supply twelve of these solar pump units, one for each village on the Kalamaro Plateau. We hope they will be of lasting benefit to your people.

Z: Our people. Yes. How do you evaluate such a benefit, James?

J: In a case like this, it is for you, the client, to decide what value to put on it.

(Azevedo is finishing a telephone call. He turns to Christine, Clarke, Davidson and Silva.)

Z: Mr. Davidson, the President has agreed to our recommendation. Your bid has been accepted. Congratulations!

(Christine, Clarke, Engel and Silva express their congratulations.)

S: Faulkner Enterprises will receive a letter of intent as a preliminary to a contract.

D: Thank you, Manuel. (Turning to Christine) I guess I ought to thank you, too, Christine. I thought for a while I was going to lose out to Pansil. But it's good to win.

J: Well, Pansil haven't exactly lost. I see they've just been awarded a contract in Peru that's twice the size of this one.

Z: Anyway, we must celebrate.

S: Yes.

Z: How long are you going to stay in Tanaku, Mr. Davidson?

D: Oh, I'm flying to San Francisco on Monday. My boss likes to be kept informed.

Z: And you, Mr. Clarke?

J: Well, I'd thought of staying on in Tanaku for a few days. I need a holiday.

D: Who doesn't?

S: Oh! Mr. Clarke, Ramilla and I have invited Christine to visit us at Yatu Beach. Would you like to come, too?

J: I'd be delighted.

S: That's a wonderful idea, Manuel.

Notes

1. The solar panels can be set up at any convenient position, ... 太阳能电池板可安装在任何方便的地方（位置）……

2. The panels are at an optimum angle, to make the most of the sunlight throughout the day. 这些太阳能电池板位于最佳的角度，能够充分利用白天的阳光。

3. What has to be done to maintain the panels? 维护太阳能电池板，我们必须做什么？

4. ... wipe the dust off the panels to keep them clean. ……擦掉电池板上的灰尘使其保持干净。

5. ... it's the way things are going. ……事情就是这么简单。

6. Perhaps the way they ought to go. 也许就应该这么简单。

7. Small sources of power, easily available. 电池规模小，容易获得。

8. This electrical motor that works the pump is controlled by a micro-computer. So, it switches itself on and off, as power is available. 驱动水泵的电动机是由微型计算机控制的，因此，

只要有电，它就能自动运转或停止。

9. The whole unit is sealed. And the whole operation is automatic. It needs no maintenance whatsoever. 整个装置是封闭的。全部操作是自动的，不需要任何维护。

10. As part of our bid for the silicon project … 作为硅产品工程项目投标的一部分……

11. … they will be of lasting benefit to your people. ……它们将会对你们的人民永久造福。

12. In a case like this, it is for you, the client, to decide what value to put on it. 在这种情况下，该是你这位客户来决定这种利益的价值。

13. Your bid has been accepted. 贵方的投标已被接受了。

14. Faulkner Enterprises will receive a letter of intent as a preliminary to a contract. 福尔克纳企业集团将收到一封意向书作为合同的初步约定。

15. I thought for a while I was going to lose out to Pansil. But it's good to win. 我曾一度认为，我们会败给潘西尔公司，但却赢得很漂亮。

16. … they've just been awarded a contract in Peru that's twice the size of this one. ……他们刚在秘鲁赢得一份大合同，是该合同的两倍。

17. My boss likes to be kept informed. 我的老板喜欢随时听到消息。

18. … I'd thought of staying on in Tanaku for a few days. ……我已想好了，在塔拉库再待几天。

19. I'd be delighted. 我很乐意（很高兴）去。

Part Five

Appendices

附录

1. Expanding Words and Phrases
（词汇拓展）

香浓可口　aromatic character and agreeable taste
香味浓郁　aromatic flavour
功效神奇　as effectively as a fairy does
身份象征　a token of status
美观耐用　attractive and durable
款式新颖　attractive designs
式样美观　attractive fashion
中国一绝　a unique of China
备有各种款式和式样　available in designs and styles
花色品种繁多　a wide range of colours and designs
花色繁多　a wide selection of colours and designs
华丽臻美　beautiful and charming
美轮美奂　beautiful and greatful
造型美观　attractive/beautiful appearance
色泽艳丽　beautiful in colour
外观色泽明亮　bright and translucent in appearance
色彩鲜艳　bright in colour
色泽光润　bright luster
科学精制/加工　by scientific process
能多次翻新　can be repeatedly remoulded
心旷神怡　carefree and joyous
用料精选　carefully-selected materials
选料考究　choice materials
清晰突出　clear and distinctive
条纹清晰　clear-cut texture
穿着舒适轻便　comfortable and easy to wear
手感舒适　comfortable feel
穿着舒适　comfortable to wear
品种齐全　complete range of articles
规格齐全　complete range of specifications
电脑验光　computer optometry
烹制方便　convenient to cook

高新技术、技术转让与国际工程合作

冬暖夏凉	cool in summer and warm in winter
防皱抗皱	crease-resistance
用户至上，顾客第一	customer first
大众所喜爱品尝的佳品	delicacies loved by all
口味鲜美	delicious in taste
性能可靠	dependable performance
称心如意	desirable
具有传统（风味）特色	distinctive for its traditional properties
款式多元新颖	diversified latest designs
方式灵活，方便客商	diversified practices and efficient services
服务上门	door-to-door service
快速干燥	drip-dry
经久耐用	durable in use / durable service
精明能干	dynamic/smart
操作简便	easy and simple to handle
业务联系方便，贸易做法灵活	easy contacts and flexible trading methods
易于润滑	easy to lubricate
维修简易	easy to repair
经济实惠	economic and practical
典雅大方	elegant and graceful
包装美观牢固	elegant and sturdy package
外观华贵	elegant appearance
气味芬芳	elegant in smell
式样雅致	elegant in style
包装美观	elegant package
外观大方	elegant shape
抗冲（强）度高	excellent in cushion effect
上等原料	excellent material
品质优良	high/fine/excellent (in) quality
技艺精湛	exquisite craftsmanship
精湛传统刺绣技艺	exquisite traditional embroidery art
做工讲究	exquisite (in) workmanship
飘逸洒脱	facile and graceful
包装新颖美观	fashionable and attractive packages
款式时髦	fashionable patterns
款式新颖	fashionable (in) style
永不褪色	fast colour
不含脂肪	fat free

暖意融融　fill the air with warmth
色泽鲜艳　fine colours
技艺精湛　fine craftsmanship
做工精细　fine (in) workmanship
结构坚固　firm in structure
芳香宜人　fragrant aroma
香味浓郁　fragrant (in) flavours
生机勃勃　full of vigor
服务周到　courteous/ full service
野味海鲜　game and seafood
赏心悦目　gladden the eye and heart
老少良伴　good companions for children as well as adults
保温性强　good heat preservation
味道纯正　good taste
种类繁多　great varieties
宾客至上　guest highest
抗热耐磨　strong resistance to heat and hard wearing
经久耐用　structural durabilities
结构坚固　sturdy construction
货源充足　sufficient supplies
老少皆宜　suitable for all
男女老少皆宜　suitable for men, women, and children
方便群众　suit the people's convenience
选材精良　superior materials
性能优越　superior performance
质量上乘　superior (in) quality
顾客至上　supremacy of customers
服务周到、礼貌　thoughtful and courteous services
保证交货及时　timely delivery guaranteed
采用先进的科学方法　to adapt advanced scientific method
采用先进的技术/工艺　to adapt advanced technology
多年使用，不出故障　to assure years of trouble-free service
荣获金/银质奖　to be awarded a gold/silver medal
荣获优质产品证书　to be awarded super-quality certificate
行销世界　to be distributed all over the world
以……著称　to be famous /distinguished for...
深受消费者欢迎和好评　to be highly praised and appreciated by consuming public
专为……设计　to be specially designed for...

高新技术、技术转让与国际工程合作

涤烦疗渴　to clear out annoyance and quench thirst
享誉中外　to enjoy high reputation at home and abroad
在国际市场上享有盛誉　to enjoy high reputation in world market
传送灵活　to ensure smooth transmission
具有独特的民族风格　to have an incomparable national style
历史悠久　to have a long historical standing
产销历史悠久　to have a long history in production and marketing
历史悠久，经验丰富，信誉可靠　to have a long history, rich experience and reliable reputation
独特的民族风格　to have a unique national style
韧硬兼蓄　to have both the quality of tenacity and hardness
助消化，除油腻　to help digest greasy food
外形永远如新　to insure a like-new appearance infinitely
促进健康　to invigorate health effectively
永葆健美　to keep you fit all the time
安心益气　to make one feel at ease and energetic
为君提供便利　to offer you the best convenience
清火明目，怡神醒脑　to produce an effect against clear vision
帮助消化　refreshment, and digestion helping
为客商提供优良服务　to provide clients with excellent service
居同类产品之冠　to rank first among similar products
减肥延寿　to reduce body weight and prolong life
满足消费者需求　to satisfy the demands of consumers
深受国内外客户信赖和赞誉　to win a high reputation and is widely trusted at home and abroad
深受青睐　to win high admiration
深受赞扬和欣赏　to win warm praise and appreciation
深受顾客欢迎　to win warm praise from customers
热带菜肴　tropical cuisine
买一送一　two for one
超值享受　unconventional enjoyment
性能无与伦比　unequal in performance
性能卓越　unique in performance
款式日新月异　up-to-date styling
先用后买　use before purchase
使用极其方便　utmost in convenience
色彩纷呈　various colours
款式多样　various fashions
款式齐全　various styles

款式活泼端庄　vivid and great in style
图案生动　vivid pattern
保暖防风　warm and windproof
防水，防震，防磁　waterproof, shock-resistant and anti-magnetic
滴水不漏　watertight
以品质优良而著称　well-known for its fine quality
举世闻名　well known in the world
品种繁多　wide varieties
久享盛誉　with a long standing reputation
久享盛名　with distinction
采用最新设备和工艺　with most up-to-date equipments and technique
沿用传统的生产方式　with the traditional methods
最新设备工艺　with updated technologies and equipments
国际金奖　world-wide gold award

2. Drills of Sentences
（句式展示）

备有塑料盒，便于安全保存。A plastic case is compartmentalized for safe storage.
本大利宽。Big capital brings big profits.
备有目录，函索即寄。Catalogues are available upon request.
备有目录，函索即寄。Catalogues will be sent upon request.
色彩夺目，迥然不俗。Colours are striking, yet not vulgar.
欢迎世界各地客商前来合作，建立和发展贸易关系。Customers from all nations are welcome to establish and develop business contacts.
价格低廉，欲购从速。Enjoy our economy price now.
竭诚欢迎客户惠顾。Enquires and orders are warmly welcome.
欢迎垂询。Enquires are cordially welcome.
无现金的社会对消费者发挥着越来越显著的影响。The cashless society is tightening its grip on consumers.
销售处资金电子过户 Eftpos（electronic funds transfer at the point of sale）
电子现金处理方式 Electronic means of processing cash
易手，换主人 Change hands
提供 eftpos 结算服务方式 Offer a cash-out eftpos service
接受或掌握电子银行的观念 Grasp the concept of electronic banking
按一下键盘，便可以把钱从一个账户上直接转到自己账户上 Directly debit money from an account at the touch of a keyboard
使……无效 Deactivate
去饭店就餐，到外边吃饭 Eat out
在饭店付账时 When it came to settling the restaurant bill
输入一个小型电子键盘 Enter into a small electronic keyboard
输入电子卡 Swipe an electronic card
（魔术师用语）您看，变！Hey presto!
现金支取服务 Cash withdrawal service
在意 Be mindful of
输入账户细目 Punch in account details
按键的瞬间便可完成付费 Pay at the push of a button
磁条码 Magnetic encoded strip
最可怕的敌人 The worst offenders
消掉磁卡信息 Counteract the magnetic information

同比增长44.4% Increase by 44.4% compared with the previous year

仅次于 Next to; only after

重要原材料商品的来源地 Source of important raw materials

全面伙伴关系和战略协作伙伴关系 Comprehensive and strategic cooperative partnership

以贸易投资便利化为主要内容 With the trade and investment facilitation as the mainstay

固定资本投资 Capital investment

扩大内需 Expand domestic demands

振兴东北工业基地 Revitalize Northeast China, a traditional industrial base

预祝……取得圆满成功 Wish ... a complete success

谈谈对单一货币的一些看法 Offer some views on the single currency

经济与货币联盟 The Economic and Monetary Union

欧元的外在价值 The Euro's external value

长期的和结构性的 Of a long-term and structural nature

欧元的短期失调 A short-term misalignment of the Euro

欧元的贬值，或美元的走强 The depreciation of the Euro or the strength of the US dollar

问题的另一个方面 The other side of the coin

无疑是得到最广泛认同的一种解释 Have been no doubt among the most popular explanations

成功地实现其首要也是最重要的目标：保持欧元区物价的稳定 Succeeded very well in delivering price stability for the Euro area, which is its primary and overriding objective

实施一些旨在增进欧元区经济发展潜力的重要措施 Implemented a number of important initiatives aiming at raising the growth potential of the Euro area

《国际货币基金组织代表团关于欧元地区经济政策的总结报告》Concluding Statement of the IMF Mission on the Economic Policies of the Euro Area

现在是欧元区基本情况最好的时期。It is hard to remember a period when the fundamentals have been as good.

与影响美国经济的不定因素相比，欧元区的不定因素似乎受到了更多的关注 Uncertainties associated with the Euro area have received more attention than those affecting the US economy

这种不平衡对欧元产生了消极的影响。This asymmetry has affected the Euro negatively

欧元失调的确给欧元区的政策制定者一些教训。The misalignment of the Euro does contain some lessons for the Euro area policy makers.

在分析中形成一个真正的欧元区整体观点 Form a true Euro area perspective in their analysis

关于欧元的前景，我们有充分的理由相信前途是乐观的 As regards to the outlook of the Euro, there is every reason to be positive

可持续性、有利于就业的经济增长 Sustainable, employment-friendly economic growth

以稳定为目标的宏观经济政策的延续 Alongside with contribution of the stability-oriented macroeconomic policies

对华投资信心 Confidence in investing in China

来华投资项目 Projects for investment in China

高新技术、技术转让与国际工程合作

有利于中国企业产业升级 Help Chinese enterprises upgrade their industries
动物源性产品 Products of animal origin
全面禁止进口 A full ban on the import of
部分解禁 Partially lift the ban on
对华实施反倾销措施 Anti-dumping measures on China
具体操作缺乏透明度 The transparency is unavailable in concrete practice.
以"中国的开放与世界的共赢"为题与各位交流 to speak to you on the topic of "Opening China and an All-win World"
万元户 thousanders; households with a ten-thousand-yuan income
带来实惠 Bring benefits to
思维上前所未有的挑战 An unprecedented challenge in the way of thinking
实现了历史的跨越 Make a historic leap
我国居民人均储蓄 The deposit per capita of the Chinese citizens
中国美国商会 American Chamber of Commerce in China
客观评价中国的市场经济建设成就,尽早给予我方完全市场经济地位 Assess impartially the achievement of China's market economy building and grant China the market economy status as early as possible
采取合作、务实和建设性的态度 Adopt a cooperative, pragmatic and constructive attitude
欧洲一体化建设 The development of the EU integration
在短期内也可能对中国造成一些经贸利益的减损 May cause s loss of economic and trade benefits to China in a short run
富有活力和长期稳定的经贸合作关系 Dynamic, long-term and stable economic and trade cooperation
跨国企业大量进入我国 Multinational corporations flow into China in large numbers
具有很大的发展空间 Have considerable space for further growth
不时出现问题和分歧 Problems and disagreement arise from time to time
积极交涉 Conduct positive negotiations, make positive representations
市场经济地位 Market economy status
对中国企业的反倾销调查存在歧视 The anti-dumping in investigation on the Chinese enterprises is of a discriminatory nature.
欧盟商会 the EU Chamber of Commerce
放松对华技术出口限制 Relax the restriction over technology export to China
充分发挥中欧技术合作的潜力 Give full play to the potentialities of Sino-EU technological cooperation
在国际事务中发挥积极作用 Play a more important role / an active role in international affairs

3. Appendices of Business Documents
商务单证附录

附录一：一般原产地证

Certificate of Origin

Original

1. Exporter	Certificate No.
2. Consignee	**CERTIFICATE OF ORIGIN** **OF** **THE PEOPLE'S REPUBLIC OF CHINA**
3. Means of transport and route	5. For certifying authority use only
4. Country/region of destination	

6. Marks and numbers	7. Number and kind of packages; description of goods	8. H. S. Code	9. Quanity	10. Number and date of invoices

11. Declaration by the exporter	12. Certification
The undersigned hereby declares that the above details and statements are correct, that all the goods were produced in China and that they comply with the Rules of Origin of the People's Republic of China.	It is hereby certified that the declaration by the export is Correct.
Place and date, signature of authorized signatory	Place and date, signature and stamp of certifying authority

附录二：

Triplicate

1. Goods consigned from (Exporter's business name, address, county) GUANGDONG KAILI TRADING CO.LTD No.108,EAST HUANSHI ROAD,GUANGZHOU, GUANGDONG,510620 CHINA			Reference No.　GZ8/10738/2023 ASEAN CHINA FREE TRADE AREA PREFERENTIAL TARIFF (Combined Declaration and Certificate) FORM E issued in THE PEOPLE'S REPUBLIC OF CHINA (COUNTRY) See Notes overleaf		
2. Goods consigned to(Consignee's name, address, country) KLOTS CABLE DISTRIBUTOR PT JT. CIDENG TIMUR 19, JAKARTA, INDONESIA					
3. Means of transport and route (as far as known) FROM SHEZHEN, CHINA TO JAKARTA, INDUNESIA Departure date　AUGUST 10, 2009 Vessel's name / Aircraft etc. DA YANG　02/025 Port of discharge　JAKARTA, INDONESIA			4. For official use ☐ Preferential Treatment Given Under ASEAN-CHINA Free Trade Area Preferential Tariff ☐ Preferential Treatment Not Given (Please state reason/s)　未审核！ Signature Of Authorized Signatory of the Importing Country.		
5. Item number	6. Marks and numbers of packages	7. Number and type of packages; description goods (including quantity where appropriate and HS number of the importing Country)	8. Origin criterion (see Notes overleaf)	9. Gross weight or other quantity	10. Number and date of invoices
2267 2268 2269	KIM'S HOME GK20090322 LOS ANGELES NO. 1—172 172 CTNS	172 CTNS PLASTIC TOY SHIPPED TO INDONESIA ***********************	"P"	6,252KGS	GK20090322 JULY 25,2009
11.Certification by the exporter The undersigned hereby declares that the above details and Statement are correct, that all the goods were produced in __CHINA__ (Country) and that they comply with the origin requirements specified for these goods in the ASEAN-CHINA Free Trade Area Preferential Tariff for the goods exported to __(INDONESIA)__ (Importing Country) __SHENZHEN, AUG. 02, 2009__ Place and date, signature and authorized signatory			12. Certification It is hereby certified, on the basis of control carried out, that the Declaration by the exporter is correct. __SHENZHEN, AUG. 02, 2009__ Place and date, signature of authorized signatory		

附录三：保险单

中保财产保险有限公司深圳分公司
The People Insurance(Property) Company of China Shenzhen Branch

18/F, Guangfa Bank Center, 83 Nonglin Xia Road
Shenzhen, P.R.C. China
510080
TEL: (86755) 87311888
FAX: (86755) 87310166

CERTIFICATE OF INSURANCE　　　**ORIGINAL**

NO: 0000123809

THE INSSURER: AIU INSURANCE COMPANY SHENZHEN BRANCH	
THE INSURED: GUANGDONG KAILI TRADING CO.LTD. No. 108, EAST HUANSHI ROAD, GUANGZHOU	AMOUNT INSURED USD20,303.00
VESSEL/CONVEYANCE: DA YANG　02/0815	SAILING ON/ABOUT JUNE 3, 2009
AT AND FROM　　YANTIAN, CHINA	TO: LOS ANGELES, USA	TRANSHIPMENT AT N/A
SUBJECT-MATTER INSURED 172 CTNS PLASTIC TOYS CONTAINTER NO.: COSU18854230　INVOICE NO.: KL090527 SAY U.S.DOLLARS TWENTY THOUSAND THREE HUNDREN AND THREE ONLY.	MARKS & NUMBERS KIM'S HOME 20090322 NEW YORK NO. 1—148
CLAIM REPRESENTATIVE THE PEOPLE INSURANCE (PROPERTY) COMPANY OF CHINA 1058 W. OLYMPIC BLDG LOS ANGELES, LA 90006 USA TEL: 1-514-627-6689 FAX: 1-514-627-6748　TALL FEE: 1 800 222879 EMAIL: USMARINECLM@AIG.COM	SETTING AGENT THE PEOPLE INSURANCE (PROPERTY) COMPANY OF CHINA 1058 W. OLYMPIC BLDG LOS ANGELES, LA 90006 USA TEL: 1-514-627-6689 FAX: 1-514-627-6748　TALL FEE: 1 800 222879 EMAIL: USMARINECLM@AIG.COM
Claim, if any, payable at: LOS ANGELES, IN US DOLLARS	Nos of ORIGINAL: THREE

INSURANCE CONDITIONS

COVERING MARINE RISKS AS PER INSTITUTE CARGO CLAUSES (A) DATED (1/1/1982)
WAR RISK AS PER INSTITUTE CARGO CLAUSES (A) DATED (1/1/1982)
(WAREHOUSE TO WAREHOUSE CLAUSE IS INCLUDED)

Subject to the conditions of Open Policy No. **MOP81033** of AIU Insurance Company Guangzhou Branch
Loss if any payable to the INSURED or order upon surrender of this Certificate.
　　It is understood and agreed that this Certificate represents and takes the place of the policy and conveys all the rights of the original policy holder (for the purpose of collecting any loss or claim), as fully as if the property is covered by a special policy direct through the holder of this Certificate, and free from any liability for unpaid premium.
　　It is agreed that, upon the payment of any loss or damage, the Insurers are to be subrogated to the extent of such payments, to all the rights of Assured under their bills of lading or other contracts of carriage.
　　It is hereby understood and agreed that in case of claim for loss under this Certificate the same shall be reported as soon as the goods are landed or the loss know, to the Agent of the Company as endorsed hereon to whom proofs of loss must in all cases be submitted for verification.
　　Claims to be adjusted according to the usage of Lloyds in Great Britain or of the ports of settlement elsewhere, but subject to the conditions of the policy and contract of insurance.
ISSURED AT: GUANGZHOU, CHINA　　　　　　　VOID IF ISSUED AFTER: **MAY 31, 2009**

DATE: **MAY 31, 2009**　　　　　　　　　　　　For and on behalf of

　　　　　　　　　　　　　　　　　　　　　　中保财产保险有限公司深圳分公司
　　　　　　　　　　　　　　　　　　　he People Insurance(Property) Company of China Shenzhen Branch

This Certificate is not valid unless countersigned by:

Amy　　　　　　　　　　　　　　　　　　　　　　　*Boney J*
─────────────　　　　　　　　　　　　　─────────────
　　　　　　　　　　　　　　　　　　　　　　　　　Authorized Representative

附录四：美国海关发票

DEPARTMENT OF THE TREASURY
UNITED STAES CUSTOMS SERVICE
19 U.S.C. 1481. 1482. 1484

SPECIAL CUSTOMS INVOICE
(Use separate invoice for purchased and non-purchased goods.)

Form Approved
O.M.B. No. 4B—R0342

1. SELLER GUANGDONG KAILI TRADING CO. LTD. No. 108, EAST HUANSHI ROAD, GUANGZHOU,GUANGDONG, 510620 CHINA	2. DOCUMENT NR. * NYP976588	3. INVOICE NR. AND DATE* KL090527 MAY 27,2009
	4. REFERENCES* N/A	
5. CONSIGNEE KIM'S HOME CENTER INC. 2940 W. OLYMPIC BLDG LOS ANGELES, LA 90006 USA	6. BUYER (*If rather than consignee*) N/A	
	7. ORIGIN OF GOODS CHINA	
8. NOTIFY PARTY* KIM'S HOME CENTER INC. 2940 W. OLYMPIC BLDG LOS ANGELES, LA 90006 USA	9. TERMS OF SALE. PAYMENT, AND DISCOUNT CIF LOS ANGELES BY SIGHT L/C 2% DISCOUNT	
10. ADDITIONAL TRANSPORTATION INFORMATION* N/A	11. CURRENCY USED USD	12. EXCH. RATE USD 1：6.55 CNY	13. DATE ORDER ACCEPTED March 20, 2009

14. MARKS AND NUMBERS ON SHIPPING PACKAGES	15. NUMBER OF PACKAGE	16. FULL DESCRIPTION OF GOODS	17. QUANTITY	UNIT PRICE		20. INVOICE TOTALS
				18. HOME MARKET	19. INVOICE	
KIM'S HOME GK20090322 LOS ANGELES NO. 1—172	60 CTNS 60 CTNS 52CTNS 172 CTNS	PLASTIC TOYS LUXURY BUS LUXURY BUS LUXURY BUS	10800PCS 10800PCS 9360PCS	CNY2.58 CNY2.72 CNY2.98	USD0.25 USD0.54 USD0.75	USD5,616.00 USD5,832.00 USD7,010.00
TOTAL:			30,960PCS			USD18,458.00

21☐ If the production of these goods involved furnishing goods or services to the Seller (e.g., assists such as dies, molds, tools, engineering work) and the value is not included in the invoice price, check box (21) and explain below	22. PACKING COST	USD500.00

27. DECLARATION OF SELLER / SHIPPER (OR AGENT)
 	
I declare:
    If there are any rebates,
    Drawback or bounties allowed (B)☐ If the goods were not sold or agreed to
(A)☐ upon the exportation of goods,   be sold, I have checked box [B] and have
    I have checked box (A) and   indicated in column 19 the price I would
    itemized separately below.   be willing to receive.
 	
    (C) SIGNATURE OF SELLER / SHIPPER (OR AGENT):
    I further declare that there is no other invoice differing
	23. OCEAN OR INTERNATIONAL FREIGHT	USD 1,520.00

	24. DOMESTIC FEIGHT CHARGE	CNY 4,500.00
	25. INSURANCE COST	USD 176.92
	26. OTHER COSTS (*Specify Below*)	N/A
28. THIS SPACE FOR CONTINUING ANSWERS N/A		

* Not necessary for U.S. Customs purposes.

Customs Form 5515 (12—20—76)

Part Five
Appendices 附录

附录五：记账单（Debit Note）

23/F, Citicorp Centre, 18 Whitfield Road,
Causeway Bay, Hong Kong
Telephone：2332 1481 Telefax：2310 2633

GUANGDONG KAILI TRADING CO., LTD.	DATE：MAY 28, 2009
No. 108, EAST HUANSHI ROAD,	OUR REF：4212730497
GUANGZHOU, GUANGDONG, 510620 CHINA	PORT OF LOADING：SHENZHEN
	PORT OF DEPARTURE：SHENZHEN
	PORT OF DISCHARGE.：LOS ANGELES, USA
ACCOUNT NO. : 64Z889	FINAL DESTINATION.：：LOS ANGELES, USA
ADDRESS NO. : 106682	BOOKING REFERENCE.：07OPS2776

NO. OF PKGS：	172 CTNS	BILL LADING NO.：	STHK802018A
WEIGHT. (KG)：	6,252.00	VESSEL：	DA YANG
CUBE. (CBM)：	58.20	VOYAGE：	02/0815
Sailing Date：	03.06.2009	Issued By：	PHOEBE LI
Consignee..... :	KIM'S HOME CENTER INC		

ORC	5.00040 USD	269.00 HKD	10531.35
Exchange Rate at	7.830000		
DOCUMENTATION CHARGES	1.000MAN HKD	150.00 HKD	150.00
HANDLING CHARGE	1.000MAN HKD	150.00 HKD	150.00
TTL RMB：11261.75			======

TOTAL AMOUNT .... HKD 10831.35
                                                            ========

E. & O. E.

"All transactions and contracts which are entered into with the company incorporate the company's trading terms and conditions. A copy of which is available on request."

Unless credit terms have been agreed by the company payment should be effected on presentation of this invoice.

Cheques should be crossed and made payable to ABX Logistics (H.K.) Ltd. Interest at 3% above prime rate will be levied on over due A/C. When payment is made by cheque no receipt will be issued unless specifically requested.

(S. 85 bills of exchange ordinance).

附录六：重量单

重量单
CERTIFICATE OF WEIGHT

发货人	发票号码
CONSIGNOR:	INV. NO.: KL090527
GUANGDONG KAILI TRADING CO.,LTD.	
No. 108, EAST HUANSHI ROAD, GUANGZHOU,	日期
GUANGDONG,510620 CHINA	DATE：JUNE 2, 2009
	信用证号码
	L/C NO.: IL080008

收货人
CONSIGNEE:　KIM'S HOME CENTER INC.
　　　　　　2940 W. OLYMPIC BLDG　LOS ANGELES, LA 90006 USA

标识及箱号 Marks and umbers	品名及规格 Article and Specification	数量 Quantity	件数 Package	毛重 G.W.	净重 N.W.
KIM'S HOME 20090322 LOS ANGELES NO. 1---148	2267 LUXURY BUS 2268 LUXURY BUS 2269 LUXURY BUS	10 800 PCS 10 800 PCS 9 360 PCS	60CTNS 60CTNS 52CTNS	1 860 KGS 2 160 KGS 1 716 KGS	2 040 KGS 2 340 KGS 1 872 KGS
	总计 TOTEL:	30,960 PCS	172CTNS	6,252KGS	5,636KGS
备注 REMARKS					

附录七：重量检验证书

中华人民共和国出入境检验检疫
ENTRY-EXIT INSPECTION AND QUARANTINE OF THE PEOPLE'S REPUBLIC OF CHINA

正本
ORIGINAL

共 1 页第 1 页 Page 1 of 1
编号 No.: 321100203088063

重量检验证书
WEIGHT CERTIFICATE

发货人
Consignor GUANGDONG KAILI TRADING CO., LTD.

收货人
Consignee KIM'S HOME CENTER INC.

品名
Commodity PLASTIC TOYS

报检数量/重量
Quantity/Weight Declared G.W. 6,252KGS;   N.W. 5,636KGS

包装种类及数量
Number and Type of Packages 172 CTNS

运输工具
Means of Conveyance S.S. DA YANG 02/0815

标记及号码
KIM'S HOME
20090322
LOS ANGELES
NO. 1—148

检验结果
Results of Inspection

The above commodity was weighted on tested scale with the result of weight as follows:
Total Gross Weight: 6,252KGS
* * * * * * * * * * * * * * * *

印章
Official Stamp 签证地点 SHENZHEN, CHINA   签证日期 Date of Issue May 20, 2009
授权签字人
Authorized Officer LI BO   签名 Signature 李波

我们以尽所知和最大能力实施上述检验，不能因我们签发本证书而免除卖方或其他方面根据合同和法律所承担的产品种类质量和其他责任。All inspections are carried out conscientiously to the best of our knowledge and ability. This certificate does not in any respect relieve the seller and other related parties from this contractual and legal obligations especially when product quality is concerned.

References
参考文献

1. 冉隆德，王恩科．贸易与经济．北京：中国海关出版社，2004.
2. 冉隆德，张兰．技术与管理．北京：中国海关出版社，2004.
3. 冉隆德，李文英．金融与投资．北京：中国海关出版社，2004.
4. ［英］科顿，［英］法尔维，［英］肯特，唐桂民．体验商务英语综合教程3．北京：高等教育出版社，2005.
5. ［英］科顿，［英］法尔维，［英］肯特，王关富．体验商务英语综合教程4．北京：高等教育出版社，2005.
6. ［英］科顿，［英］法尔维，［英］肯特，吴云娣．体验商务英语综合教程2．北京：高等教育出版社，2005.
7. 虞苏美，杨乾龙．新编商务英语口语2．北京：高等教育出版社，2004.
8. 余富林，王占斌，岳福新等．商务英语翻译：英译汉．北京：中国商务出版社，2003.
9. 邱革加，杨国俊．双赢现代商务英语谈判．北京：中国国际广播出版社，2006.
10. 吴思乐，胡秋华．世纪商务英语——谈判口语．大连：大连理工大学出版社，2007.
11. 李红英．会展英语实用教程．大连：大连理工大学出版社，2008.
12. 刘杰英．世纪商务英语——口译教程．大连：大连理工大学出版社，2008.
13. 车丽娟，贾秀海．商务英语翻译教程．北京：对外经济贸易大学出版社，2007.
14. 谢毅斌．商务英语（上，下）．北京：对外经济贸易大学出版社，2008.
15. 马玉玲等．新编商务英语实训教程·口语．北京：中国经济出版社，2008.
16. 王振光．企业生存英语．北京：清华大学出版社，2005.
17. 王逢鑫．高级汉英口译教程．北京：外文出版社，2004.
18. 刘杰英．世纪商务英语——函电与单证．大连：大连理工大学出版社，2007.
19. 钟书能．国际商务英语模拟实训教程．北京：对外经济贸易大学出版社，2009.